LONDON MATHEMATICAL SOCIETY LECTURE NOTE SERIES

Managing Editor: Professor M. Reid, Mathematics Institute,
University of Warwick, Coventry CV4 7AL, United Kingdom

The titles below are available from booksellers, or from Cambridge University Press
at www.cambridge.org/mathematics

London Mathematical Society Lecture
Note Series: 395

How Groups Grow

A V I N O A M   M A N N
Einstein Institute of Mathematics
Hebrew University of Jerusalem

CAMBRIDGE
UNIVERSITY PRESS

CAMBRIDGE
UNIVERSITY PRESS

32 Avenue of the Americas, New York NY 10013-2473, USA

Cambridge University Press is part of the University of Cambridge.

It furthers the University's mission by disseminating knowledge in the pursuit of education, learning and research at the highest international levels of excellence.

www.cambridge.org
Information on this title: www.cambridge.org/9781107657502

First published 2012

A catalogue record for this publication is available from the British Library

ISBN 978-1-107-65750-2 Paperback

Cambridge University Press has no responsibility for the persistence or accuracy of URLs for external or third-party internet websites referred to in this publication, and does not guarantee that any content on such websites is, or will remain, accurate or appropriate.

Contents

Preface

The topic of growth entered group theory, with a geometric motivation, at the middle of last century. It associates to each finitely generated group a number-theoretical function, its *growth*, and investigates the relationship between the properties of the group and of its growth function. The subject has attracted attention gradually, until, in about 1980, two seminal theorems were proved: first Misha Gromov determined the groups of polynomial growth, and a short time later Slava Grigorchuk constructed groups of intermediate growth. Both theorems were the starting points for rich mathematical developments. As far as I know, there is no detailed treatment of growth in book form. Of course, I should not ignore Pierre de la Harpe's remarkable text *Topics in Geometric Group Theory*, a large part of which is devoted to growth, but in that text most results are quoted without proofs (an exception is the chapter about Grigorchuk's group, but even about that group we were able to include in the present text more recent results).

For several years now, I have been teaching courses devoted to one or the other of the above results. Most of these were one-semester courses in the Einstein Institute of Mathematics in the Hebrew University, but some were given at several Italian universities. The present notes were prepared for, and based on, these courses. They can be divided roughly into four main parts: introductory, polynomial growth, intermediate growth, and miscellany. The first part includes the first two chapters, of which the first consists mainly of definitions and examples, and sets the context for the rest of the book. The second chapter was written mainly for the benefit of readers who are either non-specialists, or beginners, in group theory, but Section 2.2 of that chapter, which develops the elementary theory of growth, should be read carefully. Section 2.5 is a digression about isoperimetric inequalities. Then follow seven chapters devoted to polynomial growth. Chapter 3 discusses groups of linear growth, for which a completely elementary treatment is available, and indeed preceded Gromov's theorem; the rest of the book is independent of that chapter. Then come three chapters about the growth of nilpotent, soluble, and linear groups. Chapter 4 includes a classification, up to commensurability, of nilpotent groups of small growth degree, and Chapter 5 includes a reduction of the proof of the uniformity of exponential growth of soluble groups to the polycyclic case. While most of the notes are essentially self-contained, Chapter 6, on linear groups, consists mostly of results that are quoted without proof. Then we introduce asymptotic cones, and finally, in Chapter 8 we prove Gromov's theorem. The approach is similar to Gromov's original one, as modified by Wilkie and van den Dries, and Gromov himself. In the definition of asymptotic cones we were influenced by Cornelia Drutu's nice survey [Dr 02], but of course we had to supply the details. It seems that this is the first detailed treatment of asymptotic cones in a book form. A different approach to Gromov's theorem was suggested recently by Bruce Kleiner, applying harmonic functions on Cayley graphs. While this has the advantage of not relying on the deep solution of Hilbert's fifth problem, the method is very different from the ones of this book, and we refer to it only in passing. Chapter 9 derives corollaries of Gromov's theorem for groups that are infinitely generated, but are of locally polynomial growth. The readers, or instructors, who want to arrive quickly at Gromov's theorem will take what they need from the first two chapters, and skip Chapter 3 and Sections 4.2 and 5.2.

The next four chapters deal with intermediate growth. First we construct Grigorchuk's group, following Grigorchuk's original approach, applying permutations of the unit interval. Then we derive several of the many remarkable properties of that group, in particular proof that it has intermediate growth, giving explicit lower and upper bounds. In the next chapter we describe generalizations of the construction, and also other approaches, via actions on regular finitary trees, or finite automata, and some other examples of intermediate growth. The next two chapters relate intermediate growth to two other important group theoretical notions: amenability and residual finiteness. The chapter on amenability contains further discussion of isoperimetry. If one is interested mainly in intermediate growth, it is possible to go directly from Chapter 2 to this part (except that in Chapter 12 we quote from Chapter 7 the notion of ultralimit).

The last four chapters are the "miscellany" part. We first show how to calculate the growth of, say, amalgamated products, or other group-theoretical constructs, from the growth of the components. We also give a detailed calculation of the growth of one group of geometrical interest, to demonstrate that even for groups of very easy description, such calculations are far from trivial. In the next chapter we discuss the generating growth function. We want first to point out the applicability of analytic methods. Then we prove the rationality result for finitely generated abelian groups, with a proof that relates the group-theoretical growth to a ring-theoretical one. We also treat briefly, without proofs, the growth of hyperbolic groups. The next chapter proves uniform exponential growth, with explicit bounds, for several classes of groups, amalgamated free products, HNN-extensions, and groups of positive deficiency, in particular one-relator groups. The final chapter introduces and discusses briefly conjugacy class growth, a different type of growth from the previous one, but closely related to it.

We end with a list of open research problems.

There are many topics that are not touched upon in this book; to do so would have delayed the publication considerably, perhaps indefinitely. We do not discuss connections with geometry, e.g. the relation between volume growth in a Riemannian manifold and the growth of its fundamental group, or between the volume growth of a Lie group and the growth of a discrete dense subgroup. Kleiner's proof was mentioned already. As we said, we treat uniform exponential growth for several classes of groups, but the important cases of soluble or linear groups are only quoted without proofs, except for a partial result for soluble groups. We also only quote without proofs the results by Grigorchuk and John S. Wilson, that the polynomiality assumption in Gromov's theorem can be relaxed, if we restrict ourselves to residually nilpotent, or residually soluble, groups. These, and other interesting ideas, may form the contents of another publication.

Preliminary versions of these notes have been circulating for some time, and I received comments on them from quite a few people, as well as from my students, and from an anonymous referee. My sincere thanks to all of them. And thanks to Eva Goldman for drawing the figures in Chapter 11.

1

Introduction

Let G be a finitely generated group, generated by $x_1, \ldots, x_d$, say. Each element $x \in G$ can be written as a *word* in the generators, i.e. as a product $y_1 \cdots y_n$, where each y_i is either one of the generators or its inverse. The number n is called the *length* of the word. (The identity element is represented by the empty word, which has length zero.) Usually, the same element can be represented by many words. Of all of these, we choose one of minimal length (this word is not necessarily unique) and call this length the *length* $l(x)$ of x. We write $a_G(n)$ for the number of elements of length n, and $s_G(n)$ for the number of words of length at most n, i.e. $s_G(n) = \sum_{i=0}^{n} a_G(i)$. We term $a_G(n)$ and $s_G(n)$ the *growth functions* of G. More specifically, $a_G(n)$ is the *strict growth function* and $s_G(n)$ is the *cumulative growth function* of G. Our interest is in these two functions, their properties, and their relationship with the structure and properties of G. The subscript G will be often omitted, if it is clear from the context which group is meant.

Example 1 G is finite iff $a_G(n)$ is eventually 0, equivalently iff $s_G(n)$ is eventually constant. On the other hand, if G is infinite, then $a(n) > 0$ for each n, and $s(n) \geq n + 1$.

Exercise 1.1 Prove the statements just made about the growth functions of infinite groups.

Example 2 If $G = \mathbb{Z}$ is infinite cyclic, then $a(n) = 2$ for all n (except for $n = 0$; $a_G(0) = 1$ for all groups).

Example 3 Let $G = F^d$ be free of rank d. Then each element has a unique expression as a reduced word. A word of length $n + 1$ is obtained by multiplying one of length n, say $y_1 \cdots y_n$, by any generator or its inverse, except for y_n^{-1}. Thus $a(n+1) = a(n)(2d-1)$, so we have $a(0) = 1$, $a(1) = 2d$, and $a(n) = 2d(2d-1)^{n-1}$ for $n \geq 1$.

If G is any d-generator group, words of length $n+1$ are obtained from ones of length n in the same way as in this example, but they need not all represent distinct elements of G, and they may also be equal to shorter words. Thus we have:

Proposition 1.1 *If G is generated by d elements, then for $n > 0$ we have $a(n) \leq 2d(2d-1)^{n-1}$.*

Before proceeding, let us clarify our use of the notion "a generating set". We do not insist that such a set be minimal in any sense. Thus, it need not consist of the minimal possible number of generators, and it is possible that a proper subset of it is still a generating set. However, we insist that no generator is equal to another, or to the inverse of another. Indeed, if this assumption is not met, we can omit a generator that is equal to another without changing the growth functions.

Exercise 1.2 Let G have d generators and the same growth function as F^d. Then $G \cong F^d$.

Example 4 Let G be the free product of $2d$ groups of order 2, generated by $x(1), \ldots, x(2d)$. Then each element has a unique expression as a product $x(i_1) \cdots x(i_n)$, in which no two consecutive factors have the same index. It follows that the growth function is the same as in the previous example. In particular, the group $D_\infty := C_2 * C_2$ has the same growth function as $\mathbb{Z}$ (C_n denotes a cyclic group of order n). That group is known as the *infinite dihedral group*. If the two cyclic factors are generated by x and y, say, then $z = xy$ has an infinite order, generates an infinite cyclic subgroup C of index 2, and $D_\infty = CC_2$, where the last factor may be either of the two cyclic free factors.

Example 5 Take G to be the free product of d copies of $\mathbb{Z}$ and e copies of C_2. The growth function is (for $n > 0$) $a(n) = k(k-1)^{n-1}$, where $k = 2d + e$. It follows that there is no bound on the number of groups with the same growth function.

Proposition 1.2 *If the growth function of G is $a(n) = k(k-1)^{n-1}$, then G is (isomorphic to) one of the groups of the last example.*

Proof Let the generators of G be $x_1, \ldots, x_d, y_1, \ldots, y_e$, where the y_i are the generators that have order 2. Then $k = 2d + e$ (consider $a(1)$). Let H be a free product of d copies of $\mathbb{Z}$ and e copies of C_2. Then there is a homomorphism from H to G, mapping the generators of the factors $\mathbb{Z}$ in H to the x_i, and mapping the generators of order 2 in H to the y_j. Since the two groups have the same growth function, the map is 1–1, and $G \cong H$. QED

Problem 1 Can there be infinitely many groups with the same growth function?

Problem 2 What properties do groups with the same growth function have in common?

The last example shows that the growth function does not determine G. But neither does G determine its growth function, because it may have different sets of generators, a point that we have ignored so far.

Example 6 Let $G = \mathbb{Z}$, and consider the two numbers $\{2, 3\}$. This is a set of generators, since each integer can be written as $2r + 3s$. Suppose that in that expression $r \geq 3$, then we can replace it by $2(r - 3) + 3(s + 2)$, reducing the length from $r + s$ to $r + s - 1$ (if $s \geq 0$. For negative s the reduction is even bigger). Similarly, if $r \leq -3$, we replace the above expression by $2(r + 3) + 3(s - 2)$ (note that in this case the term $2r$ contributes $|r| = -r$ to the length). This implies that the minimal length is obtained when we write our integer in the form $3k, 3k + 2$, or $3(k - 1) + 2 \cdot 2$, if it is positive, the lengths being $k, k + 1$, and $k + 1$, respectively. For negative numbers the minimal expression is $3k, 3k - 2$, or $3(k+1) - 2 \cdot 2$, with lengths $|k|$ and $|k| + 1$. There is a slight exception for the integers $1 = 3 - 2$ and $-1 = 2 - 3$, the only ones for which the minimal expressions have r and s of different signs. It follows that $a(0) = 1$, $a(1) = 4$, $a(2) = 8$, and $a(n) = 6$ for $n \geq 3$. Any pair of coprime integers is a set of generators of $\mathbb{Z}$, and the growth function relative to it can be computed similarly. For example, for the generators $\{2, 5\}$ the expression of minimal length would be one of $5k, 5k + 2, 5k - 2, 5k + 4, 5k - 4$, with lengths $k, k + 1, k + 1, k + 2, k + 2$, and growth function $a(n) = 10$ (for $n \geq 3$).

Exercise 1.3 If $\mathbb{Z}$ is generated by r and s, with $0 < r < s$, then the corresponding growth function is $a(n) = 2s$, provided n is large enough.

As another illustration, consider the infinite dihedral group D_∞, and take as generators x and z, using the notation of Example 4. It is easy to see that each element can be written uniquely in the form z^k or $z^k x$, and that this is a shortest presentation. It follows that $a(1) = 3$ and $a(n) = 4$ for $n \geq 2$.

To formalize the relationship between two growth functions of the same group, we introduce the following concept.

Definition   Two functions f and g from $\mathbb{N}$ to $\mathbb{N}$, or from $\mathbb{N}$ to $\mathbb{R}$, or from $\mathbb{R}$ to $\mathbb{R}$, are *equivalent* if there exists a real positive number A such that $f(x) \leq Ag(Ax)$ and $g(x) \leq Af(Ax)$.

Proposition 1.3 *Two growth functions of the same group are equivalent.*

Proof   Let $s = s_{G,X}$ and $t = s_{G,Y}$ be two growth functions of G. Express each element of Y as a word in the elements of X, and each element of X as a word in the elements of Y, and let A be the maximal length of the resulting set of words. It is then clear that for each $x \in G$ we have $l_X(x) \leq Al_Y(x)$, and it follows that $t(n) \leq s(An)$. Similarly, $s(n) \leq t(An)$.   QED

In analogy with Problem 2, we may ask, what properties do groups with equivalent growth functions have in common? But this seems too general. Thus, if we define a function $f(n)$ to be of *exponential growth*, if there exist numbers $a, b > 1$ such that $a^n \leq f(n) \leq b^n$, then all functions of exponential growth are equivalent to each other. In particular all groups of exponential growth (i.e. groups with growth functions of exponential type) have equivalent growth functions, but, as we shall see, this class of groups includes groups of widely differing structures.

Example 7 Let $G = \mathbb{Z}^d$, a free abelian group, with the natural set of d free generators. Each element can be written uniquely in the form $x_1^{e_1} \cdots x_d^{e_d}$. Writing a_d for $a_{\mathbb{Z}^d}$, and s_d similarly, we see easily that $a_d(n) = a_{d-1}(n) + 2\sum_{k=1}^{n} a_{d-1}(n-k)$. Using the known formulas for sums of powers of the first n integers, and induction on d, we obtain that $a_d(n)$ and $s_d(n)$ are polynomials in n of degrees of $d-1$ and d, respectively. For example, $a_2(n) = 4n$ and $a_3(n) = 4n^2 + 4n + 2$ $(n > 0)$.

Example 8 Let $G = H \times K$ be a direct product. Given sets of generators of H and K, their union is a set of generators for G, and for an element $x = (u, v) \in G$ we have $l_G(x) = l_H(u) + l_K(v)$. It follows that $a_G(n) = \sum_{r=0}^{n} a_H(r)a_K(n-r)$. This equality reminds us of the multiplication rule for polynomials, or power series, and suggests the following:

Definition   The *strict generating growth function* of G is the infinite series $A_G(X) = \sum_0^\infty a(n)X^n$. We will often refer to it simply as the *generating growth function*.

The *cumulative generating growth function* of G is $S_G(X) = \sum s(n)X^n$.

Proposition 1.4

(a) $A(X) = (1 - X)S(X)$.
(b) If $G = H \times K$, and G, H, K have generating growth functions $A(X)$, $B(X)$, and $C(X)$ respectively, then $A(X) = B(X)C(X)$.

Here part (b) is just a reformulation of the formula in Example 8, and part (a) is the analytic form of the equality $a_n = s_n - s_{n-1}$. The equivalent form $s_n = a_0 + a_1 + \cdots + a_n$ is expressed analytically by the formula $S(X) = A(X)(1 + X + X^2 + \cdots)$.

The series $A_G(X)$ is sometimes written as $\sum_{a \in G} X^{l(a)}$.

Part (b) enables an alternative approach to Example 7. It also implies that given two groups H and K, the growth function of their direct product depends only on the growth functions of the factors. Thus, if two groups G and H have the same growth function, and K is any group, then $G \times K$ and $H \times K$ also have the same growth function. Moreover, the three groups $G \times G$, $G \times H$, and $H \times H$ have the same growth function. By taking direct products with more than two factors, it seems that we can again find any number of different groups with the same growth function. But we need to be careful! We said "seems" above, because it is possible for two finitely generated groups to be non-isomorphic but have isomorphic direct squares. The most popular theorem on uniqueness of direct decompositions, the Krull–Schmidt theorem, often does not apply in our context. A very general uniqueness theorem is proved in [Ku 56, section 47]. It implies, e.g., that in a group with trivial centre any two direct decompositions into directly indecomposable groups are isomorphic. It follows that if we take for G and H two non-isomorphic groups with the same growth function of the type discussed in Examples 4 and 5 above, then, with one exception, two direct products of G and H are isomorphic only if they have the same number of factors of each isomorphism type. The exception occurs when G and H are $\mathbb{Z}$ and D_∞. For that case, see the following exercise.

Exercise 1.4

(a) Show that two direct products of copies of $\mathbb{Z}$ and of D_∞ are isomorphic only if they have the same number of factors of either type.

(Hint: consider the maximal finite subgroups of the direct product, and the factor groups over the subgroup generated by all these subgroups. An alternative approach: consider the factors groups G/G^2 and G/G^3. Here G is the direct product, and G^n is the subgroup generated by all nth powers in G).

(b) Try to generalize this approach to the other pairs of groups with the same generating function provided by Examples 4 and 5.

Example 9 Let G be a semidirect product of H and K. We recall that means that $G = HK$, where $H \cap K = 1$ and $H \triangleleft G$, but K need not be normal in G. To know the structure of G we have to know not only the structure of H and K, but also the action of K on H, i.e. we have to know for each $x \in K$ the automorphism σ_x of H defined by $\sigma_x : z \to z^x$ ($z \in H$). The map $x \to \sigma_x$ is a homomorphism of K to Aut(H). If this homomorphism is trivial, we recapture the direct product, but sometimes the semidirect product is isomorphic to the direct one even if the above homomorphism is non-trivial, e.g. if $K = H$, acting on itself by inner automorphisms. Returning to the general case, each element of G can be written uniquely as xz, with $x \in K$ and $z \in H$. Given sets of generators for H and K, again their union is a set of generators for G. However, writing x and z in terms of these generators does not necessarily yield a shortest possible word for this product, even if the expressions for x and z are the shortest possible. This is so because the word $z^{x^{-1}}x$ may be shorter than the word xz, which is equal to it, when we write $z^{x^{-1}}$ in terms of the generators of H. However, this phenomenon cannot occur if the automorphisms σ_x preserve the length of words in H, equivalently if these automorphisms permute the generators of H and their inverses (the elements of length 1) among themselves. In that case we do get the shortest expression for xz by using the shortest ones for x and z, and therefore the growth function is the same as for the direct product of H and K.

Note that that yields another source for finding different groups with the same growth function. As an illustration, consider the infinite dihedral group. It can be considered as a semidirect product $\mathbb{Z}$ by C_2, the latter group acting on $\mathbb{Z}$ by inversion. Thus it has the same growth function as the direct product $\mathbb{Z} \times C_2$. On the other hand, we know already that, with a different choice of generators, it has the same growth function as $\mathbb{Z}$ (see Example 4).

Note also that the union of the sets of generators of H and K does not always yield the most natural set of generators for G. For example, the direct product of two cyclic groups of finite, relatively prime orders is itself cyclic. There are even examples of non-trivial, finitely generated groups which are isomorphic to their direct products with themselves [TJ 74]. An interesting example is obtained by considering sets of generators X and Y for groups H and K, and taking for the direct product $G = H \times K$ the set of generators consisting of $X \cup Y$ and of the set of products $\{xy \mid x \in X, y \in Y\}$. It is easy to see that the length of an element (h, k) relative to this set of generators is $\max(l_X(h), l_Y(k))$, implying $s_G(n) = s_H(n)s_K(n)$.

Exercise 1.5 Prove that, no matter which sets of generators we choose for $\mathbb{Z}$ and for $\mathbb{Z} \times C_2$, the two groups do not have the same growth function.

Example 10 Let $G = C_2 * C_3$, say $G = \langle x, y | x^2 = y^3 = 1 \rangle$. It is well known that $G \cong \mathrm{PSL}(2, \mathbb{Z})$ (for a simple proof of that, see, e.g., appendix B of [Ku 56], or Section II.28 of [Hr 00]). If some word $w \in G$ ends with x, then we get a longer word by multiplying by either y or y^{-1}. But if w ends with y, then multiplying by x yields a longer word, multiplying by y^{-1} yields a shorter word, and multiplying by y yields a word ending in $y^2 = y^{-1}$, so $l(wy) = l(w)$. The same happens if w ends in y^{-1}. Now let us perform two consecutive multiplications, and check when the length increases both times. If w ends in x, we have first to multiply by y or y^{-1}, and then by x, while if w ends in y or y^{-1}, we have to multiply first by x and then by y or by y^{-1}. In either case there are two possibilities, which means that $a_G(n+2) = 2a_G(n)$. Starting from the values $a(1) = 3$ and $a(2) = 4$, we obtain $a(2n + 1) = 3 \cdot 2^n$ and $a(2n) = 4 \cdot 2^{n-1} = 2^{n+1}$.

For a general free product we have

Proposition 1.5 *If $G = H * K$, and G, H, K have generating growth functions $A(X), B(X), C(X)$ respectively, then*

$$\frac{1}{A} - 1 = \left(\frac{1}{B} - 1 \right) + \left(\frac{1}{C} - 1 \right).$$

Proof Let $G = H * K$. Write a typical element as $x = u_1 v_1 u_2 \cdots v_r$, with $u_i \in H, v_i \in K$. Here all factors are different from the identity element, except possibly for u_1 or v_r. The number of elements of the above form of length n is $\sum a_H(s_1)a_K(t_1)a_H(s_2) \cdots a_K(t_r)$, where the sum is subject to the constraints $\sum s_i + \sum t_i = n$ and $s_i \geq 1$ for $i > 1$ and $t_i \geq 1$ for $i < r$. This is the coefficient of X^n in $B(X)C(X)((B(X) - 1)(C(X) - 1))^{r-1}$. Summing over r, we obtain

$$A(X) = \frac{B(X)C(X)}{1 - (B(X) - 1)(C(X) - 1)} = \frac{B(X)C(X)}{B(X) + C(X) - B(X)C(X)}.$$

Taking inverses we get the formula in the statement of the proposition.

QED

Again, this proposition enables an alternative approach to examples **3,4,5,** and **10.** Also, as in the case of direct products, it gives us another way of producing any number of groups with the same growth function. Unlike the case of direct products, free decompositions have strong uniqueness properties. Any finitely generated group is the free product of finitely many freely indecomposable groups, and the indecomposable factors are uniquely determined, up to isomorphism [Ku 56, section 35].

The above examples may give the impression that calculating the growth function of a group is a straightforward matter. This is not at all the case. Indeed, for most groups that calculation is impossible. The word "impossible" is used here in a very precise sense. We recall that mathematicians often equate the number theoretical functions that "can be calculated" with the so-called *recursive functions*. Unlike the phrase in inverted commas, the notion of a recursive function can be defined precisely, and the assumption that these functions are exactly the *computable* ones is known as Church's Thesis (after Alonzo Church, 1903-1995). Thus our claim that often the growth function is impossible to calculate means that for many groups this function is not recursive. We will not, however, repeat the definition of a recursive function, but will continue to use the expression "can be calculated" in an intuitive, naive way.

Let us consider groups that are not only finitely generated, but also *finitely presented*. We recall that this means that our group G is given by finitely many generators, say $x_1, \ldots, x_d$, and finitely many *relations* $w_1 = 1, \ldots, w_r = 1$, where each w_j is a word in the x_i, and G is the most general group which can be generated by d elements satisfying these equalities. In more precise terms, we consider the w_j as elements of the free group F^d, and $G \cong F^d/N$, where N is the normal closure in F^d of $w_1, \ldots, w_r$. The words w_j are called the *relators*, and the equalities $w_j = 1$ are called the *defining relations*, of G. If H is any group with d generators, which we also denote $x_1, \ldots, x_d$, then H also is isomorphic to a factor group of F^d, say $H \cong F^d/K$. The defining relations of G hold in H, i.e. $w_1 = \cdots = w_r = 1$ is true in H, if and only if $w_j \in K$, which is the same as $N \leq K$, and then $H \cong (F^d/N)/(K/N) \cong G/L$, for some $L \lhd G$. In that sense G is the "most general" group satisfying its defining relations; any other group satisfying them is isomorphic to a factor group of G.

Recall also that we say that *the word problem is soluble in G*, if there exists an algorithm to decide if two words in the generators are equal (i.e. represent the same element) in G. Since $x = y$ is the same as $xy^{-1} = 1$, it suffices to determine when a word is equal to the identity. It is easy to enumerate all the elements of F^d, say by using a lexicographic order, and thus index all words in $x_1, \ldots, x_d$ by the natural numbers. Solving the word problem is then the same as being able to tell if a particular natural number is the index of a word that is equal to 1 in G. In other words, the characteristic function of the set of such indices should be computable, or, using Church's thesis, recursive. A set whose characteristic function is recursive is itself termed a *recursive set*.

There are many groups with insoluble word problems (see [Ro 95, Ch. 12]), so our claim about the frequent incomputability of growth functions is justified by:

Proposition 1.6 *Let G be a finitely presented group. The word problem is soluble in G if and only if the growth function of G is recursive.*

Proof If the word problem is soluble, we calculate $a_G(n)$ by listing all words of length n and checking which of them are equal to each other or to shorter words. For the converse, notice first that if N is, as above, the normal closure in F^d of the set of relators, then the elements of N are exactly all the products of conjugates of the relators and their inverses, and these products can be listed, say by ordering r-tuples in F^d by the sum of r and the lengths of the r words, ordering the finitely many tuples with a given sum arbitrarily, and for each r-tuple writing down all the products of conjugates by these r elements of relators and their inverses. That means that all the words that are 1 in G are being listed one by one, possibly with repetitions, so if a certain word w in F^d is the identity in G, we will find that out by carrying out this listing process till we see w. But if $w \neq 1$, it will never appear in our list; but at any given moment we do not know if w did not appear yet because it is not 1, or just because we did not wait long enough. Thus to solve the word problem we have to know which words are not the identity. Now suppose that we are able to compute $s_G(n)$, and that w has length n (in F^d). We start by writing down all words of length n at most. Then we carry out the above process of listing the words representing 1. This is the same as listing all equalities $u = v$ between two words. From time to time we find equalities between two words of length at most n, and this reduces the number of elements of length n in G, either by showing that a word of length n has actually a shorter length in G, or by showing that two such words represent the same element. We continue this till this number goes down to the computed value of $s(n)$. At this stage we know that we are not going to find any more equalities among words of length at most n, and in particular if our word w was not shown yet to be 1, it is actually not the identity. Thus the word problem is solved.   QED

It is interesting that a similar result holds in a sort of dual situation. By a *recursive permutation* we understand a permutation of the natural numbers that is recursive as a function. These permutations form a group. Let G be a finitely generated subgroup of this group. We can list the set of pairs (w, n), where w is a word on the generators of G and n is a natural number, and since the generators are recursive, we can evaluate $w(n)$ for each such pair. If $w \neq 1$ in G, then after listing enough pairs, we will find one for which $w(n) \neq n$. That means that this time, we are going to know which words are not the identity, and to solve the word problem we have to be able to decide which words are trivial in G.

Proposition 1.7 *With the above notation, the word problem is soluble in G if and only if the growth function $a_G(n)$ is recursive.*

Proof If the word problem is soluble, the growth function is recursive, as in the previous proposition. For the reverse direction, given a word w of length n, we again list all words of length at most n, and as explained above, we test them for inequalities. At each stage of the testing we are going to discover that certain words of length at most n are not equal to each other, while about other words we will be uncertain yet. In any case, we obtain a lower bound for the number of elements of G of length at most n, and when this bound reaches $s(n)$, a computable number, we know that we have representative words for all elements of length at most n of G. We continue to test w until we find that it is different from $s(n) - 1$ of these representatives, and then we know that w is equal to the remaining representative. In particular we know if it is the identity or not.   QED

This proposition has a converse, which actually yields a sort of characterization of groups with a soluble word problem.

Proposition 1.8 *A finitely generated group G has a soluble word problem if and only if it is isomorphic to a group generated by finitely many*

recursive permutations on the natural numbers, and has a recursive growth function.

Proof By the previous proposition, we have only to prove that if G is finitely generated and has a soluble word problem, it is isomorphic to a group generated by finitely many recursive permutations. Let $x_1, \ldots, x_d$ be the generators of G. The solubility of the word problem means that, if we enumerate all the reduced words on these generators, i.e. all the elements in the free group generated by them, we have an algorithm to tell if two words are equal or not in G. We can then enumerate the elements of G as follows: for each natural number n, assume that all words of length less than n have been checked already, and as a result all elements of G of length less than n were listed, and labelled by numbers, say from 1 to k. We then multiply the words of length $n - 1$ on the right by the generators and their inverses, thus producing all words of length n. These are checked for equality to shorter words and to each other. The first new one found will be labelled by $k + 1$, the next by $k + 2$, etc. It is clear that this process results in a list of all group elements. Moreover, for each generator x_i this process evaluates the right multiplication by it, as a permutation on the group elements (or on the numbers labelling them), and thus these permutations, which generate a group isomorphic to G, are recursive.   QED

Having by now acquired by the preceding examples and discussion some feeling for our subject, let us stop and review briefly its history, at the same time explaining our aim in these notes. For all that the notion of growth seems to be a very natural one, it arose first not in pure group theory, but in geometric applications, when it was noted that the growth of the volume of balls in a manifold is closely related to the growth of its fundamental group. We are, however, interested more in the strictly group theoretical aspects, so will not discuss these applications. The main results in the subject so far are concerned with the order of magnitude of the growth functions. We saw in Proposition 1.1 that the growth is at most exponential, and in Example 7, that it can be polynomial. A major theorem of M. Gromov characterizes completely the groups of polynomial growth. These are the nilpotent-by-finite groups. That means the groups in question contain nilpotent normal subgroups with finite factor groups. Such groups are also termed *virtually nilpotent* or *almost nilpotent*, and a similar terminology is employed for other group properties as well. The proof of Gromov's theorem was achieved in several stages, each one relying on its predecessors. First it was proved for soluble groups (J. Milnor & J.A. Wolf). Then J. Tits showed it for linear groups (that means groups of matrices over some field; do not confuse them with groups of linear growth, which we will also discuss), and finally Gromov came up with a fantastic argument for the general case. We will present the case of soluble groups in full. Tits' result is an immediate corollary of a beautiful theorem of his describing linear groups, which we are able here only to quote and show how it implies the growth result. Gromov's argument produces, for a group of polynomial growth, a nice metric space on which our group acts as a group of isometries. The elaborate construction of this space is as a sort of a limit of a sequence of metric spaces, which are constructed very naturally from the given group. Following Gromov's proof, L. van den Dries and A.J. Wilkie came up with a different construction of the limit space, employing tools of non-standard analysis, in particular ultra-products. That idea was simplified by Gromov, who introduced the name *asymptotic cone* for the limit space, and we will adopt that approach.

Once the metric space and the action on it are produced, Gromov quotes the so-called "solution of Hilbert's fifth problem", to reduce the problem to the case of linear groups. The solution of Hilbert's fifth problem, by A.M. Gleason and D. Montgomery & L. Zippin, is a deep and formidable result, so again we will be content just to quote this result and point out its relevance in our context. Recently, B. Kleiner has published a proof of Gromov's theorem that does not rely on Hilbert's fifth problem ([Kl 10]; see [Ta 10] for an elementary exposition). However, the construction of asymptotic cones has proved to be very useful besides its application for Gromov's theorem.

The next question is what types of growth are possible. For groups of polynomial growth we will show that the degree, which can be defined as $lim\ sup\ \log s_n/\log n$, is an integer, which can be calculated explicitly. In the simple examples discussed above, the growth was either polynomial or strictly exponential. It took some time for examples of groups of *intermediate growth*, i.e. neither polynomial nor exponential, to be found. This was done by R.I. Grigorchuk, and some of these examples will be described below. While Grigorchuk's construction produces a continuum of possible intermediate growths, these growth functions are always of order of magnitude bigger than $2^{\sqrt{n}}$, and it is not yet known if groups whose growth is at most that bound, but not polynomial, exist.

Note that we have $s_G(m+n) \leq s_G(m)s_G(n)$. It is well known that this implies that $\omega(G) := \lim s_G(n)^{1/n}$ exists (and indeed the limit is the infimum of $\{s_n^{1/n}\}$). Groups of exponential growth can be defined by the fact that $\omega(G) > 1$. Actually, the exact value of the above limit depends not only on G, but also on the chosen set of generators (but if the inequality $\omega(G) > 1$ holds for one set of generators, it holds for all such sets). We say that the exponential growth is *uniform*, if $\Omega(G) := \inf \omega(G) > 1$, the infimum taken over all finite sets of generators of G. J.S. Wilson [Wi 04(1)] constructed examples of groups of exponential but non-uniform growth. Variations of this construction were given by L. Bartholdi [Ba 03], who has also constructed more examples of groups of intermediate growth, by Wilson himself [Wi 04(2)], and by others. D.V. Osin proved that the growth of soluble groups is either polynomial or uniformly exponential. A similar result is known for linear groups (E. Breuillard & T. Gelander, following a similar result for characteristic 0 by A. Eskin, S. Mozes & H. Oh), for hyperbolic groups, one-relator groups, groups of positive deficiency, and various other classes. A more general question than the existence of groups of non-uniform exponential growth is the following: what are the possible values of $\omega(G)$ and of $\Omega(G)$? Not much is known here.

A different type of problems has to do with the generating function $A(x)$ introduced above. This has radius of convergence $1/\omega(G)$. For abelian groups, or even abelian-by-finite, it is always a rational function. For non-abelian groups, however, $A(x)$ can be rational for one set of generators and irrational, even transcendental, for another set. This very intriguing situation, which can occur even for nilpotent groups of class two, can be interpreted as saying that the group itself, independently of any concrete representation of it, tells us which sets of generators are "nicer" than others. In one of our final chapters we will discuss these problems briefly. We have already pointed out the relationship between recursiveness of the growth function and solubility of the word problem.

To end this introductory chapter, we describe a geometrical language that is often applied while discussing growth and other notions concerning finitely generated groups. Let G be generated by the finite set X. The *Cayley graph* of the pair (G, X) has all the elements of G for vertices, and two elements are connected by an edge if, and only if, they differ by multiplication by one of the generators, i.e. elements $g, h \in G$ are connected iff $h = gx$ or $g = hx$, for some $x \in X$. For example, the Cayley graph of $\mathbb{Z}^n$, with the standard set of generators, can be identified with the *lattice points* in $\mathbb{R}^n$, i.e. all points with integer coordinates, together with the horizontal and vertical segments connecting these points. The graph of a free group, relative to a set of free generators, is a *tree*, i.e. the graph contains no closed loops. The Cayley graphs of free products of copies of $\mathbb{Z}$ and C_2 are also trees, and these are the only groups whose graphs are trees. This is so because any relation $w = 1$ in G, for some word w on the generators of G, except for relations of the type $x^2 = 1$, defines a loop containing the identity in the graph, and conversely, such a loop defines a relation. The Cayley graph can be considered as a metric space, by defining the distance $d(g, h)$ to be the length of the shortest path connecting g and h (the graph is connected, because X generates G). The elements of length n in G are exactly the vertices at distance n from the identity element; more generally we have $d(g, h) = l(g^{-1}h)$. Note that if $a \in G$, then left multiplication by a, i.e. the map $g \to ag$, is an isometry of the graph. It follows that our metric space is *homogeneous* (also called *transitive*), i.e. for any two points g, h in it there exists an isometry mapping g to h. Mapping each element a to the corresponding isometry embeds G in the group of isometries of the space. This space and the action of G on it are the starting point in Gromov's proof.

2

Some Group Theory

In this chapter we recall some elementary group theoretical results. The reader who is reasonably familiar with group theory may skip, or skim briefly, most of this chapter, but we advise reading carefully Section 2.2, where we develop some basic properties of growth.

2.1 Finite Index Subgroups

Proposition 2.1 *A finite index subgroup of a finitely generated group is itself finitely generated.*

Proof Let $G = \langle x_1, \ldots, x_d \rangle$, let $|G : H| = s$, and let $\{a_1 = 1, \ldots, a_s\}$ be a *transversal* for H in G, i.e. a set of representatives for the right cosets of H. Let $x = y_1 \cdots y_n$ be any element of G, where each y_i is one of the generators or their inverses. Let a_{i_1} be the representative of Hy_1, and write $x = y_1 a_{i_1}^{-1} a_{i_1} y_2 \cdots y_n$, then let a_{i_2} be the representative of $Ha_{i_1}y_2$, and write $x = y_1 a_{i_1}^{-1} \cdot a_{i_1} y_2 a_{i_2}^{-1} \cdot a_{i_2} y_3 \cdots y_n$, etc., finally obtaining $x = y_1 a_{i_1}^{-1} \cdot a_{i_1} y_2 a_{i_2}^{-1} \cdot \cdots \cdot a_{i_{n-1}} y_n a_{i_n}^{-1} \cdot a_{i_n}$. Here the terms $a_j y_k a_l^{-1}$ lie in H, so if $x \in H$, we have $a_{i_n} = 1$, and x is written as a product of the finitely many *triple products* $a_j y_k a_l^{-1}$. QED

Remark Since the elements a_j and y_k determine the element a_l such that $a_j y_k a_l^{-1} \in H$, the proof shows that if we denote by $m(X)$ the minimal number of generators of the group X, then $m(H) \leq 2m(G)|G : H|$. But $a_j x_k^{-1} a_l^{-1} = (a_l x_k a_j^{-1})^{-1}$, therefore it suffices to take the triple products in which y_k is one of the x_i, thus cutting $m(H)$ by half. A further reduction can be made by ordering all the words on the x_i lexicographically (assuming that the x_i precede the x_i^{-1}, say), and choosing each a_i as the first element in its coset under this ordering. Then, if

$a_i = z_1 \cdots z_r$, then $z_1 \cdots z_{r-1}$ is also an a_j, and $a_j z_r a_i^{-1} = 1$, so this element can be omitted from our set of generators. This applies to all a_i save $a_1 = 1$, so we omit a further $s - 1$ generators, the final inequality being

$$m(H) - 1 \leq (m(G) - 1)|G : H|.$$

This inequality is called *Schreier's inequality*. It is the best possible. We will show this by proving that if G is free, then equality holds. Since it requires only a little more effort, we will prove also the fundamental property of subgroups of free groups, given in the following result.

Theorem 2.2 (Nielsen–Schreier Theorem) *A subgroup of a free group is free.*

Proof Let F be a free group freely generated by elements x_i (whose number need not be finite or even countable). Fix some well-ordering of the generators and their inverses, and then order all elements of F, first by their length as words in the generators, and elements of the same length are ordered lexicographically. This is a well-ordering of F. Let H be a subgroup of F, and, as above, choose as representative of each coset the element which is first by our ordering. As we saw, this guarantees that an initial segment of one of the representatives is itself a representative. Letting $\{a_j\}$ be the set of these representatives, we already know that H is generated by its non-identity elements of H of the form $a_j x_k a_l^{-1}$. We call this last element $y_{j,k}$, and we want to show that these elements are free generators of H.

Step 1. The elements $a_j x_k a_l^{-1}$ are reduced as written.

Otherwise, there is some cancellation in that word, and this is possible only if the x_k in the middle is cancelled. Suppose that it is cancelled by the letter on its left. That means that a_j has the form $u x_k^{-1}$, where u is an initial word of a_j, and u is also the representative of its coset. But $u = a_j x_k$, and thus u is the representative of $H a_j x_k$, i.e. $u = a_l$, and $a_j x_k a_l^{-1} = u x_k^{-1} x_k a_l^{-1} = 1$. If x_k is cancelled by the letter on its right, we can make a similar argument, or apply the same argument to the word $a_l x_k^{-1} a_j^{-1}$, the inverse of our word.

Step 2. If x is any reduced word in the non-identity elements $y_{j,k}$, then, when writing x as a word in x_i, the elements x_k (or their inverses) in the middle of $y_{j,k}$, are not cancelled.

Suppose that some x_k is cancelled, let us say by the y to its right, say in a product $y_{j,k} y_{r,s}$. Then the a_l^{-1} occurring in $y_{j,k}$ must be cancelled completely first, and then x_k is cancelled. One way for this to happen is for a_r to start with $a_l x_k^{-1}$. Then this last element, as an initial segment of a representative, is itself the representative of its coset. But looking at $y_{j,k}^{-1} = a_l x_k^{-1} a_j^{-1}$ we see that a_j represents the coset of $a_l x_k^{-1}$. That means that $a_j = a_l x_k^{-1}$, and thus $y_{j,k} = 1$.

If x_k is cancelled by the x_s in $y_{r,s}^{-1}$ to its right, then $y_{r,s}^{-1}$ must start with a_l and $x_s = x_k$, but that means that $y_{r,s} = y_{j,k}$, and the word in the ys is not reduced. Finally, it may be that x_k is cancelled by a letter in the a_s^{-1} that ends $y_{r,s}$. Then inverting x we get a word in which $y_{r,s}^{-1} y_{j,k}^{-1}$ occurs, and the x_s^{-1} in $y_{r,s}$ is cancelled in the same manner as in the first case for x_k above, and, as there, this implies that $y_{r,s} = 1$. In the same way we deal with the possibility that x_k is cancelled on the left, and that ends Step 2.

It is clear that the last step shows that H is free on the indicated generators. Moreover, if $|G : H|$ is finite, the count of generators above shows that Schreier's inequality is an equality. QED

Proposition 2.3 *A finitely generated group contains only finitely many subgroups of a given finite index.*

Proof Let again G be generated by $x_1, \ldots, x_d$ and $|G : H| = s$. Given any $r < s$, and any r cosets $Ha_1, \ldots, Ha_r$, their union K is not all of G, therefore K cannot be closed for multiplication on the right by the generators and their inverses, i.e. for some i and some generator x_j either $Ha_i x_j$ or $Ha_i x_j^{-1}$ is a new coset. It follows that each coset has a representative of length $s - 1$ or less, and the triple products above have length at most $2s - 1$. Since there are only finitely many elements of such lengths, there are only finitely many ways to choose the generators of H. QED

Corollary 2.4 *If $|G : H| = s$ is finite, then H contains a finite index subgroup K that is normal in G, and if G is finitely generated, we can take K to be characteristic in G.*

Proof Let K be the intersection of all the conjugates of H, or, if G is finitely generated, the intersection of all subgroups of G of index s.

Remark One can get from the proof of Proposition **2.3** an estimate for the number of subgroups of G of index s, but this bound turns out to be very crude. The best possible bound is obtained by noting that a subgroup of index s determines a representation of G as a transitive permutation group, of degree s, on the cosets of H. This representation associates to an element $x \in G$ the permutation $Ha \to Hax$ of the cosets. If we denote the cosets by the numerals 1 to s, with H corresponding to 1, then H is determined as the stabilizer of 1. Since the cosets distinct from H can be numbered arbitrarily from 2 to s, the number of subgroups of index s is equal to the number of homomorphisms of G onto transitive subgroups of the symmetric group S_s, divided by $(s-1)!$. If $G = F^m$ is free of rank m, the number of homomorphisms of G into S_s is $(s!)^m$, yielding an upper bound $s \cdot (s!)^{m-1}$ for the number of index s subgroups of F^m. Some further analysis, taking into account the non-transitive homomorphisms, yields an exact recursive expression for the number of subgroups, and in particular shows that the number of index s subgroups in F^m is asymptotic, as $s \to \infty$, to the above upper bound. See [LS 03] for details.   QED

2.2 Growth

Our main interest in these notes is in questions related to the order of magnitude of growth functions, and in that connection we are now going to define some terms.

First, note that by writing a word of length $m + n$ as a product of one of length m and one of length n, we get that $a(m + n) \leq a(m)a(n)$. Therefore $\omega(G) := \lim_{n \to \infty} a(n)^{1/n}$ exists and is finite. For similar reasons, $s(G) := \lim s(n)^{1/n}$ exists, and it is clear that $s(G) \geq \omega(G)$. We assume that G is infinite, and then $a(n) \geq 1$ for all n, and therefore $\omega(G) \geq 1$. Given any $\epsilon > 0$, we have $a(n) \leq (\omega(G) + \epsilon)^n$, if n is large enough. Therefore $s(n) \leq A + n(\omega(G) + \epsilon)^n$, for some constant A, and it follows that $s(G) \leq \omega(G)$. Thus $\omega(G) = s(G)$ and in the sequel we will use only the notation $\omega(G)$ for this invariant. We know already from Proposition 1.1 that if G is generated by d elements, then $\omega(G) \leq 2d - 1$, and we have just seen that $\omega(G) \geq 1$. The exact value of $\omega(G)$ depends not only on G, but also on the set of generators X, and if that dependence is important, we will use the notation $\omega_X(G)$. As we will see in a moment, whether $\omega(G) = 1$ or not does not depend on X. This justifies part of the following

Definition

(a) G has *exponential growth* if $\omega(G) > 1$, and *subexponential growth*, if $\omega(G) = 1$. The number $\omega(G)$ is called the *exponential growth rate* of G (rather, of (G, X), where X is the relevant generating set of

G). Sometimes the number $\log \omega(G)$ is called the *entropy* of G (or of (G, X)).

(b) G has *uniform exponential growth* if $\inf_X \omega_X(G) > 1$. We denote this infimum by $\Omega(G)$, and call it the *minimal growth rate* of G (and $\log \Omega(G)$ is termed the *minimal entropy*, or the *algebraic entropy* of G) (if G is finite, we put $\Omega(G) = 0$).

(c) G has *polynomial growth*, if there exist numbers c and s such that $s_G(n) \leq cn^s$, for all n. If $s = 1$ or 2, say, we say that G has *linear*, or *quadratic*, growth, etc.

(d) If G has polynomial growth, its *degree* is defined by $d(G) = \inf(s \mid$ and there exists c such that $s_G(n) \leq cn^s) = \limsup_{n \to \infty} \frac{\log s(n)}{\log n}$.

(e) G has *intermediate growth*, if its growth is neither exponential nor polynomial.

Clearly, groups of polynomial growth are of subexponential growth. Later we will construct many groups of intermediate growth.

Proposition 2.5

(a) *The type of growth of G, i.e. exponential, intermediate, or polynomial, does not depend on the choice of generators. If the growth is polynomial, then the degree does not depend on the generators.*

(b) *If $H \leq G$ and $N \triangleleft G$, with both G and H finitely generated, and if G is of subexponential or polynomial growth, then so are H and G/N. If the growth is polynomial, then the degrees of H and G/N do not exceed the degree of G.*

(c) *If $|G : H|$ is finite, then G and H have equivalent growth functions, and in particular have the same type of growth, and if this growth is polynomial, then G and H have the same degree. If N is finite, then G and G/N have equivalent growth functions, and $\Omega(G) = \Omega(G/N)$. In particular, G and G/N have the same type of growth, and if one of them is uniformly exponential, so is the other. If the growth is polynomial, then G and G/N have the same degree.*

(d) *If G has polynomial growth, $|G : H|$ is infinite, and H is finitely generated, then $d(H) \leq d(G) - 1$, and if N is infinite and finitely generated, then $d(G/N) \leq d(G) - 1$.*

(e) *If $|G : H|$ is finite, and H has uniform exponential growth, so does G.*

(f) *Free (non-cyclic) groups have uniform exponential growth. If G is k-free (i.e. all k-generated subgroups of G are free), $k \geq 2$, and G is not cyclic, then G has uniform exponential growth.*

Proof Let $H := \langle y_1, \ldots, y_e \rangle$, with $y_i \in G$, and let $k = \max l(y_i)$. Then $s_H(n) \leq s_G(kn)$. This implies one half of (b), and (a) is the special case $H = G$. The claim in (b) about G/N is obvious, if we choose as generators for G/N the cosets of the generators of G. Moreover, if H is of finite index, let r be the maximal length of the elements in a system of representatives for the cosets of H. Given an element in G of length at most n, write it as xu, where $x \in H$ and u belongs to our system of representatives. Then $k := l(x) \leq n + r$. Write $x = y_1 \cdots y_k$, where each y_j is either a generator or an inverse of a generator. Then $x = y_1 u_1^{-1} \cdot u_1 y_2 u_2^{-1} \cdot u_2 y_2 \cdots y_k u_k^{-1}$, for some u_j in our system of representatives. This shows that relative to the generators of the form $u_i^{-1} y_j u_m$ of H, we have $l_H(x) \leq k$, and therefore $s_G(n) \leq |G : H| s_H(n + r) \leq |G : H| s_H((r + 1)n)$. This proves one part of (c). For the other, we noted already that a generating system X of G maps on one, say Y, of G/N. Moreover, any generating system Y of G/N is the image of a generating system X of G. Simply take preimages of the elements of Y and supplement them by generators of N. An element of G of length n (relative to X) maps on one of length at most n of G/N, and exactly $|N|$ elements of G map on each element of G/N, therefore $s_{G/N}(n) \leq s_G(n) \leq |N| s_{G/N}(n)$. It follows that $\omega_X(G) = \omega_Y(G/N)$, that $\Omega(G) = \Omega(G/N)$, and that if the growth is polynomial, then $d(G/N) = d(G)$.

Now assume that $|G : H| = \infty$. Let $X = \{x_1, \ldots, x_m\}$ be a set of generators of G, which we take to contain also a set of generators for H, and let $Hu_1, \ldots, Hu_n$ be distinct cosets of H. The set $K := Hu_1 \cup \cdots \cup Hu_n$ is not closed under multiplication on the right by the generators of G and their inverses, otherwise $K = G$. Thus one of the elements $u_i x_j^{\pm 1}$ represents a new coset. Starting with $u_1 = x_1$, this argument shows that for each n we can find n distinct cosets of H which are represented by elements of length at most n, so we may assume that $l(u_i) \leq i$. If $z_1, \ldots, z_k$ are elements of length at most n of H, then the elements $z_i u_j$ are all different from each other, hence $s_G(2n) \geq n s_H(n)$, so if the growth is polynomial, then $d(G) \geq d(H) + 1$.

Next, let N be an infinite, finitely generated normal subgroup of G, the latter still assumed to be of polynomial growth. Let X be a finite set of generators of G, containing a set of generators Y for N. It is then clear that $s_{G,X}(2n) \geq s_{G/N, XN/N}(n) s_{N,Y}(n) \geq n s_{G/N}(n)$, implying that $d(G) \geq d(G/N) - 1$.

Let $|G : H| = s$, and suppose that H has uniform exponential growth. That means that there exist constants $A > 0$ and $c > 1$ such that $s_{H,Y}(n) \geq Ac^n$, for all n and all sets of generators Y of H. Here A

may depend on Y, but c does not. Let $X = \{x_1, \ldots, x_m\}$ be any set of generators of G. We saw in the proof of Proposition 2.3 that there exist representatives for the cosets of H of length at most $s-1$, and the proof of Proposition 2.1 shows that H can be generated by a set Y of elements whose X-length is at most $2s - 1$. Thus $s_{G,X}((2s - 1)n) \geq s_{H,Y}(n) \geq A(c^{1/(2s-1)})^{(2s-1)n}$, so $\omega_X(G) \geq c^{1/(2s-1)}$ and $\Omega(G) \geq \Omega(H)^{1/(2s-1)}$.

Now let $G = F^d$ be free and non-abelian, and let X be a set of generators of G. Then two of these generators, say x and y, do not commute. The subgroup $H = \langle x, y \rangle$ is then free, non-abelian, and 2-generated, and it is known that then the generators x and y of H are free generators, therefore $a_n(G) \geq a_n(H) \geq 4 \cdot 3^{n-1}$, so $\Omega(G) \geq 3$.

A more elaborate argument shows that actually $\Omega(G) = 2d - 1$. We already know that if X is a set of free generators, then $\omega_{G,X} = 2d - 1$, and therefore $\Omega(G) \leq 2d - 1$. Let X be any finite set of generators of G. Recall that $G/G' \cong \mathbb{Z}^d$, a free abelian group of rank d. We can embed $\mathbb{Z}^d$ in $V := \mathbb{Q}^d$, a d-dimensional vector space over the rational field $\mathbb{Q}$. Let Y be the set of images of X in $\mathbb{Z}^d$. Then Y is a set of generators for V, and therefore contains a basis. The subgroup (not subspace) K of V that is spanned by that basis is contained in $\mathbb{Z}^d$, and is isomorphic to it. Let $T \subseteq X$ be a set of d preimages of the elements of that basis. The subgroup $H = \langle T \rangle$ of G is free, and in the mapping of G onto G/G', H maps onto K, and therefore H has rank d. Since T is a subset of X, we obtain $\omega_{G,X} \geq \omega_{H,T} = 2d - 1$. Thus $\Omega(G) = 2d - 1$.

Finally, let G be k-free. Then any two elements of G generate a free subgroup. If G is abelian, that means that any 2-generated subgroup is cyclic, and an immediate induction shows that each finitely generated subgroup is cyclic, and G itself is cyclic. Thus we assume that G is not abelian. Then any set of generators S contains two non-commuting elements, say x and y, which generate a free subgroup F of rank 2, and $\omega(G, S) \geq \omega(F, \{x, y\}) \geq \Omega(F) = 3$, implying $\Omega(G) \geq 3$. If $k > 2$ we can improve that. We may assume that G is not free. Among the subsets of S that generate free subgroups of rank less than k, let T be maximal. Then $T \neq S$, because G is not free. Let $x \in S - T$. Then $H := \langle T \rangle$ can be generated by at most $k - 1$ elements (free generators), therefore $F = \langle H, x \rangle$ is free, and the maximality of T shows that F has rank k. Then $\omega(G, S) \geq \omega(F, \{T, x\}) \geq \Omega(F) = 2k - 1$. Therefore $\Omega(G) \geq 2k - 1$ (this argument is taken from [SW 92]; examples of non-free k-free groups are constructed in [AO 96], [Bu 01]). QED

It seems to be unknown whether a finite index subgroup of a group of uniformly exponential growth group is itself of uniform exponential growth. We remark also that the assumption in (**d**) that H and N are finitely generated is superfluous; this follows from the structure theorem for groups of polynomial growth. We do not know if there is an elementary direct proof.

Exercise 2.1 If $G = H \times K$ has polynomial growth, then $d(G) \leq d(H) + d(K)$. Equality holds if $K \cong H$.

Exercise 2.2 Let $G = H \times K$. Show that $\Omega(G) = max(\Omega(H), \Omega(K))$.

Exercise 2.3 Let G have polynomial growth, let $N \triangleleft G$, and assume that N contains an infinite finitely generated subgroup. Show that $d(G/N) \leq d(G) - 1$.

Exercise 2.4 Let $G = H * K$. Let X and Y be sets of generators for G and H, respectively, and generate G by $X \cup Y$. Then $\omega_{G,X\cup Y} \geq \omega_{H,X} + \omega_{K,Y}$.

Exercise 2.5 Let $|G : H| = 2$. Show that $\Omega(G) \geq \Omega(H)^{0.4}$.

Exercise 2.6 Let G have exponential growth. We say that *the growth rate of G is realized*, if G has a set of generators S such that $\omega(G, S) = \Omega(G)$. Suppose that $\Omega(G) \geq \Omega(H)$. Prove that the growth rate of $G \times H$ is realized, if and only if either $\Omega(G) > \Omega(H)$ and the growth rate of G is realized, or $\Omega(G) = \Omega(H)$, and the growth rates of both G and H are realized.

Remark It follows from Gromov's theorem that equality holds in Exercise 2.1 always.

Problem If G has uniform exponential growth, and H is a finite index subgroup of G, does H have uniform exponential growth?

To widen the scope of applicability of Proposition 2.5, we introduce the following

Definition Two groups G and H are *commensurable* if they contain isomorphic finite index subgroups, i.e. if there exist $K \leq G$ and $L \leq H$ such that $|G : K|$ and $|H : L|$ are finite, and $K \cong L$. And G and H are *quasi-commensurable* if there exist subgroups $M \triangleleft K \leq G$ and $N \triangleleft L \leq H$, such that $|G : K|$, $|H : L|$, $|M|$, and $|N|$ are all finite, and $K/M \cong L/N$.

Commensurability and quasi-commensurability are equivalence relations, and it follows from the last proposition that quasi-commensurable groups have the same type of growth. More than that is true.

Proposition 2.6 *Two quasi-commensurable groups have equivalent growth functions.*

To prove the proposition, it suffices to show that if H is a finite index subgroup of G, and N is a finite normal subgroup of G, then G, H, and G/N have equivalent growth functions. All the necessary inequalities were obtained already during the proof of Proposition 2.5.

In the other direction, we note that in the examples that we have given of two, or more, groups with the same growth function, these groups were often commensurable. For example, if we have, as in Example 9, a semidirect product HK, with K acting on H by permuting the set X of the generators and their inverses among themselves, then the action of K is determined by the permutations it induces on the finite set X, therefore it induces on H a finite group of automorphisms. Let $L = C_K(H)$. Then $|K : L|$ is finite, and G contains the direct product $H \times L$ as a subgroup of finite index, and thus G is commensurable with the direct product $H \times K$, which has the same growth function as G.

Another type of group with the same growth function involves free product (Examples 4 and 5), and to analyze it we recall the Kurosh subgroup theorem [Ro 96, Th. 6.3.1]. According to that theorem, a subgroup H of a free product G is itself a free product, where one free factor is a free group (possibly trivial), and the others are the intersections of H with certain conjugates of the free factors of G. We apply that to a free product G of d copies of $\mathbb{Z}$ and e copies of C_2, as in Example 5. There is a natural homomorphism of G onto the direct product of e copies of C_2. The kernel H of that homomorphism has a finite index, and intersects the free factors isomorphic to C_2 trivially, because these factors map injectively onto their image. Since H is normal, it intersects also the conjugates of these factors trivially. It follows that H is a free product of a free group and groups isomorphic to $\mathbb{Z}$ (the conjugates of subgroups of $\mathbb{Z}$). Thus H itself is free. Moreover, H is non-abelian, unless $d = 0$ and $e = 2$, i.e. unless G is infinite dihedral. This follows from the formula given for the rank of H in [Ro 96, 6.3.1]. Another way to see this is to note that any non-trivial free product, other than $C_2 * C_2$, has exponential growth (prove this!), and thus it cannot have a cyclic finite index subgroup. Thus G is commensurable with a non-abelian free group, and it follows from Schreier's formula that any two non-abelian free groups of finite rank are commensurable.

It is also clear that commensurability is preserved under direct products. But it need not be preserved under free products. In [Hr 00, Section IV.46] there are examples of commensurable groups G and H such that the free squares $G * G$ and $H * H$ are not quasi-commensurable.

A more general equivalence relation which still preserves growth up to equivalence is obtained by looking at the Cayley graph as a metric space.

Definition

(a) A map $f : S \to T$ between two metric spaces is *quasi-isometry*, if there exists a positive constant C, such that $(1/C)d(x,y) - C \leq d(f(x), f(y)) \leq Cd(x,y) + C$. Here $d(-,-)$ denotes the distance in either space, and $x, y \in S$.

(b) The metric spaces S and T are *quasi-isometric*, if there exists a quasi-isometry f from S to T, and a constant C, such that each point of T is at a distance at most C from some point of $f(S)$.

(c) Two finitely generated groups are *quasi-isometric*, if their Cayley graphs are quasi-isometric.

It is easy to see that quasi-isometry is an equivalence relation on metric spaces (prove this!). If G is a finitely generated group, the Cayley graphs of G with respect to two different (finite) generating sets are quasi-isometric, therefore we did not have to specify the generating sets in (c) above, and quasi-isometry is an equivalence relation on finitely generated groups.

Proposition 2.7 *Two quasi-commensurable groups are quasi-isometric.*

For the proof it suffices to show that any group is quasi-isometric to subgroups of finite index, and also to factor groups over finite normal subgroups. The verification of this, which is very similar to the proof of Proposition 2.5, is left for the reader, as well as the verification that the growth functions of two quasi-isometric groups are equivalent.

The examples mentioned just preceding the last definition show that quasi-isometric groups need not be quasi-commensurable. Further examples are given in Sections 30, 47 and 48 of Chapter IV of [Hr 00].

2.3 Soluble and Polycyclic Groups

Definition A group G is *soluble* if it has a normal series

(1) $$1 = G_n \lhd G_{n-1} \lhd \cdots \lhd G_1 = G.$$

with abelian factor groups G_i/G_{i-1}.

Subgroups and factor groups of soluble groups are soluble, as are extensions of one soluble group by another.

Recall that the *commutator* $[x, y]$ of two elements x, y is defined by $xy = yx[x, y]$, i.e. $[x, y] = x^{-1}y^{-1}xy$, and that the *commutator subgroup* (also termed the *derived subgroup*) is the subgroup of G generated by all the commutators. It is denoted by G' or $[G, G]$. More generally, given any two subgroups H and K of G, we write $[H, K]$ for the subgroup generated by all commutators $[x, y]$ with $x \in H$, $y \in K$.

We write $G^{(i)} = [G^{(i-1)}, G^{(i-1)}]$, with $G^{(0)} = G$. Then $G^{(1)} = G' \lhd G$, and G/G' is abelian. Moreover, if $N \lhd G$, with G/N abelian, then $N \geq G'$, so that G' is the smallest normal subgroup of G with an abelian factor group. It follows from this, or directly from the definition, that G' is a characteristic subgroup, and so are the subgroups $G^{(i)}$, the terms of the *derived series* of G. It also follows that if G is soluble, with the series G_i above testifying to it, then $G^{(i)} \leq G_i$, so that $G^{(n)} = 1$. Thus a group is soluble if and only if its derived series reaches the identity at some stage. Moreover, if n is the smallest index such that $G^{(n)} = 1$, then all series with abelian factors have length at least n, and we say that G has *derived length* (or *solubility length*) n.

Groups of length 2 are often called *metabelian*.

Now recall that the following three conditions on a group G are equivalent:

(1) *Maximum condition.* Any collection of subgroups of G has a maximal element.

(2) *Ascending chain condition.* A properly increasing chain of subgroups

$$G_1 < G_2 < \cdots < G_n < \cdots$$

is finite.

(3) Each subgroup of G is finitely generated.

Groups satisfying these conditions are sometimes called *Noetherian* (after Emmy Noether, 1882–1935). Being Noetherian is inherited by subgroups, factor groups, and extensions. The following is an example of a finitely generated soluble group that is not Noetherian.

Example   Let A be the group of all linear functions $f(x) = ax+b$, with a and b rational and $a \neq 0$. The mapping $f \to a$ is a homomorphism of A onto the multiplicative group of non-zero rationals, with a kernel, consisting of all translations $x + b$, isomorphic to the additive group of rationals. Thus A is metabelian. Let $G < A$ be generated by the two functions $f(x) = 2x$ and $g(x) = x + 1$. Then $f^{-n}gf^n(x) = x + \frac{1}{2^n}$, so that the group of all translations contained in G is isomorphic to the additive group of all rationals with denominator a power of 2, and this subgroup is not finitely generated (prove this).

Proposition 2.10 *A soluble group is Noetherian if and only if it has a normal series with cyclic factors.*

Proof   If in the series (1) all factors are cyclic, then G is obtained by a series of successive extensions by cyclic groups, so it is Noetherian, because so are the cyclic groups.

   Conversely, if G is soluble and Noetherian, then in the series (1) all the factors are finitely generated abelian groups, hence direct sums of cyclic groups, so that we can refine (1) to a normal series with cyclic factors.   QED

   Because of Proposition 2.10, soluble Noetherian groups are called *polycyclic*.

Proposition 2.11 *Let G be a polycyclic group. Then any two normal series of G with cyclic factors contain the same number of infinite factors.*

Proof   Any two normal series have isomorphic refinements, therefore it suffices to consider a series as (1) and a refinement of it. Such a refinement can be achieved in a sequence of steps, each step inserting just one more term in (1): say we insert H between G_i and G_{i-1}. Here, if G_i/G_{i-1} is finite, we did not touch the infinite factors at all. But if that factor is infinite, it is isomorphic to $\mathbb{Z}$, and then $H/G_{i-1} \cong \mathbb{Z}$ and G_i/H is finite, so the number of infinite factors is unchanged.   QED

Definition   Let G be polycyclic, and let (1) be a normal series of G with cyclic factors. The number of infinite factors in this series is called the *Hirsch length* of G, denoted $h(G)$ (after Kurt A. Hirsch, 1906–1986).

   The following property of polycyclic groups is often useful.

Theorem 2.12 *A polycyclic group contains a torsion-free group of finite index.*

Proof By induction on the Hirsch length. Let (1) be a series with cyclic factors. Suppose that G_i/G_{i+1} is the first infinite factor. Then $|G : G_i|$ is finite, and it suffices to show that G_i has a torsion-free subgroup of finite index. Thus, replacing G by G_i, we may assume that G has a normal subgroup H with G/H infinite cyclic. By induction, H contains a torsion-free subgroup K of finite index, and by Corollary 2.4, we may assume that $K \lhd G$. Let G/H be generated by the coset xH. Then xK has an infinite order in G/K, therefore the subgroup $L := \langle K, x \rangle$ is an extension of the torsion-free group K by an infinite cyclic group. Thus L itself is torsion-free, and $G = LH$ implies that $|G : L| = |H : H \cap L| \leq |H : K|$ is finite. QED

We should note that polycyclic groups are a minority among finitely generated soluble groups. By that we mean that there are only countably many polycyclic groups (up to isomorphism), while there are $2^{\aleph_0}$ finitely generated soluble groups. The first claim follows from the fact that polycyclic groups are finitely presented. This follows easily by induction from the fact that extensions of finitely presented groups by finitely presented groups are themselves finitely presented. Thus polycyclic-by-finite groups are also finitely presented. There are only countably many finitely presented groups, because there are only countably many finite presentations. A continuum of finitely generated soluble groups is constructed, e.g., in [Ha 54].

2.4 Nilpotent Groups

Notation We call $Z(G)$ the *centre* of the group G. If $x, y \in G$, then $x^y = y^{-1}xy$.

Definition A group G is *nilpotent* if it has a normal series (1) such that $G_i \lhd G$ for each i, and $G_i/G_{i+1} \leq Z(G/G_{i+1})$. Such a series is termed a *central series*.

Definition The *lower central series* $\gamma_i(G)$ of a group G is defined by $\gamma_1(G) = G$ and $\gamma_{i+1}(G) = [\gamma_i(G), G]$.

It is easy to check that the elements of the lower central series are characteristic in G, and that if (1) is a central series for G, then $\gamma_i(G) \leq G_i$. Therefore G is nilpotent iff $\gamma_{c+1}(G) = 1$ for some c. Moreover, if c is the first index for which this occurs, then c is the shortest length of all central series of G, and we say that G has *nilpotency class* (or just *class*)

c, denoted $\mathrm{cl}(G) = c$. In particular, the abelian groups are the nilpotent groups of class 1.

Definition The *upper central series* $Z_i(G)$ of a group G is defined by $Z_1(G) = Z(G)$ and $Z_{i+1}(G)/Z_i(G) = Z(G/Z_i(G))$.

It is easy to check that the elements of the upper central series are characteristic in G, and that if (1) is a central series for G, then $Z_i(G) \geq G_{n-i}$. Therefore G is nilpotent iff $Z_c(G) = G$ for some c. Moreover, if c is the first index for which this occurs, then c is the shortest length of all central series of G, and therefore $c = \mathrm{cl}(G)$.

Subgroups, factor groups, and direct products of nilpotent groups are nilpotent, but an extension of one nilpotent group by another need not be nilpotent, as we see already by looking at the smallest non-abelian group, the symmetric group S_3 on three letters. The same example shows that soluble groups need not be nilpotent, while it is clear from the definition that nilpotent groups are soluble. It is well known that finite p-groups are nilpotent. To present some infinite examples, let R be any commutative ring, and let $U := U(n, R)$ be the group of *upper unitriangular n-by-n matrices*, i.e. the upper triangular matrices with 1 on the main diagonal. Let U_i be the set of matrices in U in which the first $i-1$ diagonals above the main diagonal vanish ($U_1 = U$). It is then easy to see that U_i is a central series for U, showing that U is nilpotent, and that U_i/U_{i+1} is isomorphic to the direct sum of $n - i$ copies of the additive group of R. If R is a field, then $\{U_i\}$ is actually both the upper and the lower central series. We can also consider the group T of all triangular matrices, i.e. allowing the diagonal terms to be any invertible elements of R. Then $U \triangleleft T$ and T/U is abelian, and so T is a soluble group, which is usually not nilpotent.

The following *commutator identities* hold in all groups, but they are particularly useful for nilpotent groups. They can be verified by direct evaluation.

$$(2) \qquad [xy, z] = [x, z]^y [y, z], \quad [x, yz] = [x, z][x, y]^z.$$

In particular, if $z \in Z_2(G)$, then the commutators of z are central, therefore $[xy, z] = [x, z][y, z]$. The same equality holds if we assume instead that $x, y \in Z_2(G)$.

Before stating the next identity we introduce more notation.

Notation A *simple commutator w of weight c*, denoted $w = [x_1, \ldots, x_c]$, is defined by $[x_1, \ldots, x_c] = [[x_1, \ldots, x_{c-1}], x_c]$. We sometimes say that w

is a commutator *in*, or *on*, the elements x_i. These elements themselves are considered as commutators of weight 1. We use similar notation for subgroups: $[A, B, C]$ stands for $[[A, B], C]$ etc.

The identity that we referred to is

$$[x, y^{-1}, z]^y [y, z^{-1}, x]^z [z, x^{-1}, y]^x = 1.$$

This is sometimes called the *Hall–Witt identity*. It is analogous to the Jacobi identity in Lie algebras. It implies the following useful lemmas:

Lemma 2.13 (Three subgroups lemma) *Let H, K, and L be normal subgroups of the group G. Then*

$$[H, K, L] \le [K, L, H][L, H, K].$$

From this we derive, by easy induction

$$[\gamma_i(G), \gamma_j(G)] \le \gamma_{i+j}(G).$$

Lemma 2.14 *Let $H, K \triangleleft G$, and let $H = \langle X \rangle$, $K = \langle Y \rangle$. Then $[H, K]$ is the normal closure of the set $[x, y]$, where $x \in X, y \in Y$.*

Proof It is clear that $[H, K] \triangleleft G$, so if we let N be the normal closure in question, then $N \le [H, K]$. Consider G/N. In this factor group, the generators of H (rather, of HN/N) commute with the generators of K, therefore all elements of H commute with all elements of K, so all commutators $[x, y]$ lie in N, and $[H, K] \le N$.   QED

Lemma 2.15 $\gamma_i(G)$ *is generated by all the simple commutators of weight i.*

Proof Proceeding by induction, we may suppose the claim true for i. The previous lemma shows that $\gamma_{i+1}(G)$ is the normal closure of the set of simple commutators of weight $i + 1$. Since a conjugate of a simple commutator is a simple commutator of the same weight, our lemma follows.   QED

Lemma 2.16 *Let $G = \langle X \rangle$. Then $\gamma_i(G)$ is the normal closure of all simple commutators of weight i on the elements of X.*

Proof We use induction, the case $i = 2$ being a special case of Lemma 2.14. Assuming the result for i, let N be the normal closure of all the simple commutators of weight $i + 1$ on X. If w is a simple commutator on X of weight i, then $[w, x] \in N$ for all $x \in X$, which shows that in G/N the element wN is central. Then so is $w^y N$, for any $y \in G$, i.e.

$[w^y, x] \in N$, for all $x \in X$. The induction hypothesis and Lemma 2.14 now imply that $\gamma_{i+1}(G) \leq N$.   QED

Corollary 2.17 *Let G and X be as before. Then $\gamma_i(G)/\gamma_{i+1}(G)$ is generated by the (images of) the simple commutators of weight i on X.*

Proof By the previous result, our factor group is generated by the conjugates of the commutators of weight i, but in $G/\gamma_{i+1}(G)$ commutators of weight i are central and equal to their conjugates.   QED

In particular, if X is finite, all the factor groups $\gamma_i(G)/\gamma_{i+1}(G)$ are finitely generated abelian. This implies immediately the following

Theorem 2.18 *A finitely generated nilpotent group is Noetherian (hence polycyclic).*

When discussing the growth of nilpotent groups, it will be convenient (though not essential) to avoid elements of finite order (*torsion elements*; if all elements of G have a finite order, then G is a *torsion group*; a group without non-identity torsion elements is *torsion-free*). The next few results are aimed at achieving that.

Proposition 2.19 *A finitely generated torsion nilpotent group is finite.*

Proof If our group G is abelian, the claim is clear. Otherwise, $G/\gamma_c(G)$ is finite, by induction on $c = \mathrm{cl}(G)$, and $\gamma_c(G)$ is finite, being a finitely generated abelian torsion group. Hence G is finite.   QED

Lemma 2.20 *Let G be nilpotent of class $c > 1$, let $x \in G$, and write $H = \langle x, G' \rangle$. Then $\mathrm{cl}(H) < c$.*

Proof H is generated by x and commutators. Then H' is generated normally by the commutators of these generators, which lie in $[G', G] = \gamma_3(G)$, and an easy induction shows that $\gamma_i(H) \leq \gamma_{i+1}(G)$ for all i, implying that $\gamma_c(H) = 1$.   QED

Proposition 2.21 *A nilpotent group that is generated by torsion elements is a torsion group.*

Proof It suffices to show that if x and y have finite order, so does xy. All conjugates of x have the same order as x and lie in $\langle x, G' \rangle$, so by Lemma 2.20 the normal closure $\langle x \rangle^G$ of x has smaller class than G, and by induction on that class $\langle x \rangle^G$ is a torsion group. Similarly the normal closure of y is a torsion group, and the product of these two normal subgroups is a torsion group containing xy.   QED

It follows that in any nilpotent G the torsion elements form a subgroup, the so-called *torsion subgroup* of G. We denote it by $\delta(G)$. If G is finitely generated, then $\delta(G)$ is finite, by Theorem 2.18 and Proposition 2.19. We define a series $\delta_i(G)$ of characteristic subgroups by

$$\delta_i(G)/\gamma_i(G) = \delta(G/\gamma_i(G)).$$

Proposition 2.22 *Let G be a torsion-free nilpotent group. Let $x, y \in G$ satisfy $x^n = y^n$, for some integer $n \neq 0$. Then $x = y$.*

Proof By induction on $\mathrm{cl}(G) = c$, starting from the obvious case $c = 1$. Let $c > 1$. Then y commutes with $y^n = x^n$, and thus $x^n = y^{-1}x^ny = (y^{-1}xy)^n$. But $\langle x, y^{-1}xy \rangle = \langle x, [x, y] \rangle$, and by Lemma 2.20 the last group has class less than c. The induction implies $x = y^{-1}xy$, i.e. x and y commute, so the claim follows from the case $c = 1$. QED

Proposition 2.23 *Let G be nilpotent. Then $[\delta_i(G), \delta_j(G)] \leq \delta_{i+j}(G)$. In particular, $\{\delta_i(G)\}$ is a central series.*

Proof We may assume that $\delta_{i+j}(G) = 1$. Then $\gamma_{i+j}(G) = 1$, therefore $\delta(G) = \delta_{i+j}(G) = 1$, and so G is torsion-free. Let $x \in \delta_i(G)$ and $y \in \delta_j(G)$. Then $x^n \in \gamma_i(G)$ and $y^m \in \gamma_j(G)$ for some positive n, m, therefore $[x^n, y^m] = 1$. Then $x^n = y^{-m}x^ny^m = (y^{-m}xy^m)^n$, and by the previous proposition $x = y^{-m}xy^m$, i.e. $[x, y^m] = 1$, and now a similar reasoning yields $[x, y] = 1$. QED

We now recall that a group G is *residually finite*, if the intersection of all finite index of it is the identity. By Corollary 2.4, this is the same as assuming that the intersection of all finite index normal subgroups is trivial, and this is the same as assuming that G is a subgroup of a cartesian product of finite groups. The following useful fact will be needed, e.g., in Chapter 13.

Theorem 2.24 *Polycyclic groups are residually finite.*

Proof Let G be polycyclic. We apply induction on the Hirsch length of G. By Theorem 2.12, G contains a torsion-free subgroup H of finite index. Let $A \neq 1$ be a normal abelian subgroup subgroup of H. Then A is free abelian of finite rank. For any n, the Hirsch length of H/A^n is smaller than that of G, therefore H/A^n is residually finite (recall that A^n is the subgroup of nth powers in A). Since $\bigcap_n A^n = 1$, H itself is residually finite, and since finite index subgroups of H have finite index in G, the group G is also residually finite. QED

2.5 Isoperimetric Inequalities

In this section, which is somewhat of a digression, we prove isoperimetric inequalities for groups. Recall that the *isoperimetric problem* is to find, among all plane figures of a given perimeter, one of maximal area.[1] The answer is, of course, a circle. That can be stated in the following form: given a plane figure of perimeter p, area S, and diameter d, the *isoperimetric inequality* $S \leq \frac{1}{4}dp$ holds. We first explain what is an isoperimetric inequality for graphs.

Given any graph Γ, and $A \subseteq \Gamma$, the *boundary* ∂A of A is the set of vertices at distance 1 from A, i.e. the vertices that lie outside A, but there is an edge between them and one of the vertices of A. We will also refer to ∂A as the *outer boundary* of A, while the *inner boundary* is the outer boundary of the complement of A, i.e. the vertices of A that are connected by an edge to some vertex outside A. We consider only *locally finite* graphs, i.e, graphs in which each vertex is connected by edges to only finitely many other vertices. In such a graph, the boundary of a finite set is also finite, and an isoperimetric inequality is an inequality that bounds the size of a finite set in terms of the size of its boundary.

Now let G be a finitely generated group, with growth functions $a(n)$ and $s(n)$. It is a natural assumption that if G is infinite, then $a(n) \leq a(n+1)$. Surprisingly, this is not always the case (see Section VI.A.9 of [Hr 00]). However, the inequality $s(n) \leq (n+1)a(n)$, which would have been trivial if the previous inequality were always valid, does hold [Z 00]. In the Cayley graph of G, let $B(n)$ be the *ball of radius* n around the identity, i.e. the set of elements of length at most n. Then $\partial(B(n))$ is the set of elements of length $n+1$, while the inner boundary consists of elements of length n, though not necessarily of all of them. Therefore the claimed inequality (which is equivalent to $s(n-1) \leq na(n)$) is a type of isoperimetric inequality. Its proof, which requires the notion of amenability for groups, will be given later (see Theorem 12.14). For now we will derive a weaker statement.

Theorem 2.25 *Let G be a finitely generated infinite group. Then $s(n) \leq (2n+1)a(n)$ and $s(n) \leq (2n+1)a(n+1)$.*

It is convenient to prove a more general result on certain graphs, and then apply it to the Cayley graph of G.

Given any graph Γ, recall that a *path* between the vertices x and y is

---

[1] The earliest known discussion of this problem, due to Zenodorus, probably dates to the 2nd century BC.

a chain of vertices $x_0 = x, x_1, \ldots, x_k = y$, such that for each $0 \le i < k$ there is an edge between x_i and x_{i+1}. The *length* of the path is k, and the *distance* $d(x, y)$ between x and y is defined to be the least length of a path from x to y (if x and y are not connected, the distance may be taken to be infinite, but our main interest is in connected graphs. Such a graph is a metric space under the above definition of distance). In the Cayley graph of G, $d(x, e) = l(x)$. The *diameter* of a subset A of Γ is the maximum distance between two vertices in A (or infinity, if these distances are unbounded). A *geodesic* is a path, say between x and y, whose length equals $d(x, y)$. If x_i and x_j are two points on a geodesic, it is clear that the segment of the geodesic between x_i and x_j is also a geodesic. It is also clear that there are geodesics through each vertex of Γ, and if Γ is infinite and connected, these geodesics can be of any (finite) length.

Finally, recall that Γ is *transitive*, if for any two vertices of Γ there exists an automorphism mapping one to the other.

Theorem 2.26 ([BS 92]) *Let Γ be a locally connected finite transitive graph, let A be a finite subset of Γ, and let d be the diameter of A. Assume that Γ contains vertices at distance $d + 1$, and let ∂A be either the outer or the inner boundary of A. Then $|A| \le (d+1)|\partial A|$.*

Proof Let s be the number of geodesics of length $d+1$ through a vertex x. By transitivity, s does not depend on x. Consider some geodesic L of that length through a point of A. Since the distance between the end points is $d + 1$, at least one of the points of L is outside A, and there exist a pair of vertices x, y on L, with an edge between them, such that $x \in A$ and $y \notin A$. Then x belongs to the inner boundary and y to the outer one. Thus all geodesics of length $d + 1$ through points of A intersect both boundaries, and letting ∂A be either boundary, the number of the relevant geodesics is at most $s|\partial A|$. On the other hand, through each point of A pass s geodesics, and if L is one of them, then L contains $d + 2$ points, at least one of which lies outside A, and therefore L was counted at most $d + 1$ times. Thus the number of geodesics is at least $s|A|/(d + 1)$, and $s|A|/(d + 1) \le s|\partial A|$, which is the desired inequality. QED

The remarks preceding Theorem 2.25 show that it follows from Theorem 2.26.

Another isoperimetric inequality was proved in [CSC 93]. Its formulation involves the inverse function of the growth, which we define first.

Definition   Let G be a finitely generated group. The *inverse growth function*, $\sigma_G(x)$, of G, is defined by

$$\sigma_G(x) = \min\{n \mid s_G(n) \geq x\}.$$

Theorem 2.27 *Let G be a finitely generated group, with a set S of k generators, and let A be any finite subset of G. Then*

$$|A| \leq 8k\sigma(2|A|)|\partial A|.$$

Note that this inequality bounds the size of A, on the left-hand side, by means of an expression, on the right-hand side, involving the same size! Therefore it is better to look at it not as an upper bound for $|A|$, but rather as a lower bound for $|\partial A|$. Before giving the proof, we compare this inequality with the previous one. Clearly $\sigma(s_G(n)) \geq n$. Therefore, if we take $A = B(n)$ in the theorem, the result is weaker than Theorem 2.23. On the other hand, if A is a small set with a large diameter, then Theorem 2.25 is stronger than Theorem 2.23. Moreover, the former refers to all finite subsets, not only balls. Indeed, it applies information about balls to deduce information about all finite subsets. In other words, it applies global information about G – its growth rate – compared to the local information about the subsets – their diameter – that occurred in the earlier result.

The proof of Theorem 2.27, while elementary, uses concepts and terminology that are inspired by analytic concepts. We write $x \sim y$ to denote that the vertices x and y are connected by an edge in the Cayley graph, i.e. $y = xx_i^{\pm 1}$, where x_i is one of the generators. We are interested in real functions with finite support on Γ. Let $f : \Gamma \to \mathbb{R}$ be such a function. Thus $f(x) \neq 0$ only for finitely many vertices of Γ. We define the *Norm* $\|f\|$ of f by $\|f\| = \sum_{x \in \Gamma} |f(x)|$, and the *gradient* of f is the function $(\nabla f)(x) = \sum_{y \sim x} |f(x) - f(y)|$. Then $\|\nabla f\| = \sum_{x \sim y} |f(x) - f(y)|$. We write $f_n(x) = \frac{1}{s(n)} \sum_{d(y,x) \leq n} f(y)$, and first want to estimate the difference between f and its averaging function f_n.

Lemma 2.28 $\|f - f_n\| \leq n\|\nabla f\|$.

Proof   Write $f^y(x) = f(xy)$. For each generator x_i, $\|f - f^{x_i}\| \leq \|\nabla f\|$, and if $l(y) \leq n$ then applying that inequality repeatedly yields $\|f -$

$f^y|| \leq n||\nabla f||$. Thus

$$||f - f_n|| \leq \frac{1}{s_G(n)} \sum_{x \in G} \sum_{l(y) \leq n} |f(x) - f(xy)|$$

$$= \frac{1}{s_G(n)} \sum_{l(y) \leq n} ||f - f^y|| \leq n||\nabla f||.$$

QED

Next we prove:

Proposition 2.29 *For G and f as above, and a real number c,*

$$|\{x \in G \mid |f(x)| \geq c\}| \leq \frac{2}{c}\sigma(\frac{2}{c}||f||)||\nabla f||.$$

Proof Take $n = \sigma(\frac{2}{c}||f||)$. Then $s_n \geq \frac{2}{c}||f||$, or $||f|| \leq \frac{c}{2}s_n$. If $x \in G$, then $|f_n(x)| \leq \frac{||f||}{s_n} \leq \frac{c}{2}$. If $|f(x)| \geq c$, then $|f(x) - f_n(x)| \geq \frac{c}{2}$. The elements such that $|f(x) - f_n(x)| \geq \frac{c}{2}$ contribute at least $\frac{c}{2}$ to $||f - f_n||$, therefore their number does not exceed $||f - f_n||\frac{2}{c}$. Combined with the lemma, we obtain $|\{x \in G \mid |f(x)| \geq c\}| \leq ||f - f_n||\frac{2}{c} \leq \frac{2}{c}n||\nabla f|| = \frac{2}{c}\sigma(\frac{2}{c}||f||)||\nabla f||$. QED

Now take f as the characteristic function of some finite set A, and $c = 1$. Then $|\{x \in G \mid |f(x)| \geq 1\}| = |A| = ||f||$, and $f(x) - f(y) \neq 0$ only if one of x, y is in A and the other in ∂A. Since G has k generators, each element of ∂A is connected to at most $2k$ elements, and this element can be either x or y, and thus $||\nabla f|| = \sum_{x,y \in G} |f(x) - f(y)| \leq 2 \cdot 2k|\partial A|$. Substituting these values in Proposition 2.29 results in Theorem 2.27.

Just as for Theorem 2.26, there is a version of Theorem 2.27 that holds for more general graphs than Cayley graphs. We indicate it briefly. Given a group G, a set of generators S, and a subgroup H, the *Schreier graph* associated to them has all the right cosets of H in G for vertices, with two cosets connected by an edge if one can be obtained from the other by right multiplication by one of the generators. This is a connected transitive graph, locally finite iff S is finite. In the latter case one can define growth, inverse growth, etc., as for groups, and one can prove a result analogous to Theorem 2.27.

Exercise 2.7 Formulate and prove a result analogous to Theorem 2.27, but mentioning the inner, rather than the outer, boundary.

Exercise 2.8 Formulate and prove the result about Schreier graphs that was just alluded to.

3
Groups of Linear Growth

3.1 Linear Growth

Theorem 3.1 *A group of linear growth is cyclic-by-finite.*

This special case of the characterization of groups of polynomial growth was proved by J. Justin [Ju 71], and a similar proof was later given independently by A.J. Wilkie & L. van den Dries [VdDW 84(2)]. We will present the proof, which is by a completely elementary combinatorial argument, and derive more properties of groups of linear growth. For these further properties we need to develop more group theory, and thus digress from our main line of study, i.e. growth. The results in this chapter will hardly be needed later, and the impatient reader may, if he or she wishes, skip to the next chapter, either now or following the proof of Theorem 3.4.

We start with a finite set $X = \{x_i\}$, and consider the free monoid generated freely by X, i.e. the set of all words on the alphabet X, with multiplication defined by juxtaposition (the empty word is the identity element). A word $w = y_1 \cdots y_n$, where each y_i is some x_j, is said to have *length* n, written $l(w) = n$; we say w is *periodic* with period p, or *p-periodic*, if $n > p$, and $y_i = y_{i+p}$, for $i = 1, \ldots, n - p$. A *subword* of w is a word of the form $y_i \cdots y_j$, where $1 \leq i \leq j \leq n$.

Lemma 3.2 *Let the p-periodic word w have a q-periodic subword of length at least $p + q - 1$, where $1 \leq q \leq p$. Then w is d-periodic, for $d = \gcd(p, q)$.*

Proof Intuitively, we can move the subword by multiples of p, until its translates cover most of w. Then, starting from any letter, we move it by a multiple of p into one of these translates, then we can move it by multiples of q. By this means we can move our letter by any linear combination of p and q, and in particular by d, and reach an equal letter.

Let us give a formal proof.

Let $v = y_i \cdots y_j$ be the q-periodic subword. Note that $l(v) = j - i + 1 \geq p + q - 1$, so $i - 1 + p \leq j$. Write $p = tq + r$, $0 \leq r < q$. At first we assume that q divides p, i.e. $r = 0$ and $d = q$. If $i > 1$, then $y_{i-1} = y_{i-1+p} = y_{(i-1+p)-(t-1)q} = y_{i-1+q}$. That means that the subword $y_{i-1} \cdots y_j$ is also q-periodic, and continuing in the same way we obtain that $y_1 \cdots y_j$ is q-periodic, and then we apply a similar calculation to the right of y_j to show that w is q-periodic.

The general case is proved by induction on q. There remains the case $r > 0$. In that case we have $y_k = y_{k+p} = y_{k+p-tq} = y_{k+r}$, for $i \leq k \leq j - p$, so that the subword $u = y_i \cdots y_{j-p+r}$ is r-periodic and of length $j - p + r - i + 1 \geq q + r - 1$. By induction, v has period $\gcd(q, r) = d$. Now replacing q by d, the previous paragraph shows that w is d-periodic.   QED

Theorem 3.3 *Let the word w of length $n \geq c+m$ have at most c distinct subwords of length m, where $c \leq m$. Then $w = utv$, where t is p-periodic and u and v have lengths at most $c - p$, for some $0 < p \leq c$.*

Proof Write $w = y_1 \cdots y_n$, and let $v_i = y_i \cdots y_{i+m-1}$ be the subword of length m starting at y_i. By assumption, $c + 1 + (m - 1) \leq n$, so the subwords $v_1, v_2, \ldots, v_{c+1}$ exist, and two of them are equal, say $v_a = v_b$ $(1 \leq a < b \leq c + 1)$, and then the word $y_a \cdots y_b \cdots y_{b+m-1}$ is p-periodic, for $p = b - a \leq c \leq m$. This word has length $b + m - 1 - (a - 1) = p + m$. Thus:

There exist indices $p \leq c$ and p-periodic subwords of w of length at least $p + m$.

Among the subwords satisfying this for some p, we pick one, t, of maximal length. We choose the notation so that p is the minimal period of t, and write $t = y_i \cdots y_j$ and $w = utv$. Here $l(t) = j - i + 1 \geq p + m$, implying $i + p + m - 1 \leq j \leq n$.

To prove our theorem, we will show that the lengths of u and v are at most $c - p$, and by symmetry, it suffices to establish this for u, i.e. we want to show that $i - 1 \leq c - p$.

Suppose that the last inequality does not hold. Then $i + p > c + 1$, so we can consider the subwords $v_{i+p-(c+1)}, \ldots, v_{i+p-1}$, of which two must be equal, say $v_k = v_l$ $(i + p - (c + 1) \leq k < l \leq i + p - 1)$. Then $s = y_k \cdots y_l \cdots y_{k+r+m-1}$ is an r-periodic subword of t, for $r = l - k$. Let q be the minimal period of s, and let $d = \gcd(p, q)$. Suppose that $k \geq i$.

Then, as $l \leq i + p - 1$, so that $l + m \leq j$ and $q \leq r = l - k \leq l - i < p$, the word $y_k \cdots y_{k+q+m-1}$ is a q-periodic subword of s and of t, of length $q + m$. Since $m \geq c \geq p$, we have $q + m > q + p - 1$, so by Lemma 3.2 t is d-periodic. Since $k \geq i$, it follows that $d \leq q \leq r = l - k \leq l - i \leq i + p - 1 - i = p - 1$. That contradicts the minimality of p as a period of t.

Thus $k < i$. Then s is no longer a subword of t, but $k + q + m - 1 \leq l + m - 1 \leq i + p - 1 + m - 1 < j$, and thus t and s overlap in the word $y_i \cdots y_{k+q+m-1}$, of length $k+q+m-1-(i-1) \geq (i+p-(c+1))+q+m-i = p+q+(m-c)-1 \geq p+q-1$, so if $q < p$, then again Lemma 3.2 shows that t is d-periodic, contradicting the minimality of p, while if $q > p$, we get in the same way that s is d-periodic and contradict the minimality of q. Therefore $p = q$. But if $p = q$, then the word $y_k \cdots y_j$ is p-periodic and longer than t. This final contradiction establishes the desired inequality $i - 1 \leq c - p$ and ends the proof.   QED

Theorem 3.4 *Let the group G satisfy $a_G(m) \leq m$, for some $m \geq 1$. Then G has a cyclic subgroup of finite index, and G has linear growth.*

Theorem 3.1 follows from Theorem 3.4: actually the following sharper result holds.

Corollary 3.5 *If G satisfies $s_G(m) \leq m(m+1)/2$, for some m, then G has a cyclic subgroup of finite index, and G has linear growth.*

Proof of 3.4. If G is generated by the finite set X, let us fix some order on the set $X \cup X^{-1}$, and order the words on X by length, with words of the same length ordered lexicographically. For each element $x \in G$, let the minimal word, according to that order, representing x, be called a distinguished word. If w is distinguished, and u is a subword of w, it follows that u is also distinguished. The assumption implies that all long enough distinguished words w have the form $w = utv$, as in Theorem 3.3. Here the periodic word t can be written as $t = s^k r$, for some exponent k and some words s, of length p, and r, of length at most p, where $p \leq m$. Thus $w = us^k rv$, with u, s, and rv, of length at most m. If u, s, rv, and $n = l(w)$ are given, then k and w are determined. Thus $a_G(n)$, for large enough n, is bounded by the constant $s_G(m)^3$, and G has linear growth. Of course we may assume that G is infinite. Then infinitely many elements are represented by words w which, when written in the form $us^k rv$, have the same s. That means that s, considered as an element of G, has infinite order. Then $H = \langle s \rangle$ has linear growth, so if $|G : H| = \infty$ then Proposition 2.5(d) shows that the growth of G is at least quadratic. This contradiction shows that $|G : H|$ is finite, and ends the proof. QED

Corollary 3.6 *The groups of subquadratic growth are either finite, so of growth degree 0, or commensurable with $\mathbb{Z}$ and of degree 1. All groups of linear growth constitute one commensurability class.*

The reader may wonder why groups that are quasi-commensurable with $\mathbb{Z}$ are not mentioned here. However, if G is quasi-commensurable with $\mathbb{Z}$, then G contains two subgroups $N \leq H \leq G$, with N finite and normal, H of finite index, and H/N infinite cyclic. Let xN generate H/N, and write $K = \langle x \rangle$. Then $H = KN$, therefore K has a finite index in H, and also in G. Thus groups that are quasi-commensurable with $\mathbb{Z}$ are already commensurable with it. This follows also from Theorem 3.1, which may seem an unnecessarily sophisticated proof, but it proves also that groups quasi-isometric with $\mathbb{Z}$ are commensurable with it.

The following result may be considered as dual to Theorem 3.1.

Theorem 3.7 *A group G of linear growth has a finite normal subgroup N, such that G/N is either infinite cyclic or the infinite dihedral group.*

Corollary 3.8 *A torsion-free group of linear growth is isomorphic to $\mathbb{Z}$.*

To prove Theorem 3.7 we need the concept of *transfer*. Let G be any group, and let H be a subgroup of finite index. Write $|G : H| = k$, and let $a_1, \ldots, a_k$ be representatives for the right cosets of H in G. If $x \in G$, then for each index i we have $Ha_ix = Ha_j$, for some j, and so $a_ixa_j^{-1} \in H$. We wish to consider all the elements $a_ixa_j^{-1}$ at once, and so we look at their product. However, this product is not well defined, because it depends on the order in which we pick the k cosets, and no order is preferred over the others. Therefore we consider not the product itself, but its image in H/H', an image which is well defined, because the order of the factors is immaterial in the abelian group H/H'.

Definition The mapping $T : x \to \Pi a_ixa_j^{-1}H'$ from G to H/H' is called the *transfer* of G to H/H'.

Proposition 3.9 *The transfer map is a homomorphism.*

Proof Let $x, y \in G$, let, as above, $Ha_ix = Ha_j$, and let $Ha_jy = Ha_u$. Note that $i \to j$ and $j \to u$ are permutations of the indices $\{1, \ldots, k\}$, whose product is the permutation $i \to u$ corresponding to xy, and that $a_ixa_j^{-1} \cdot a_jya_u^{-1} = a_ixya_u^{-1} \in H$. The last equation is actually a set of

k equations, one for each i, and multiplying them together shows that $T(x)T(y) = T(xy)$.   QED

Theorem 3.10 (I. Schur) *If $|G : Z(G)|$ is finite, then G' is finite.*

Proof   Write $|G : Z(G)| = k$, and consider the transfer $T : G \to Z(G)$. Consider also the permutation representation of G on the cosets of $Z(G)$, and single out one cycle in the permutation π induced by the element x. For simplicity, let us assume that that cycle is $(1, 2, \ldots, r)$. Then the contribution of that cycle to the product defining the transfer is $a_1 x a_2^{-1} \cdot a_2 x a_3^{-1} \cdots a_r x a_1^{-1} = a_1 x^r a_1^{-1} = x^r$, the last equality holding because the product, like its factors, lies in $Z(G)$. Multiplying over all cycles of π, we get $T(x) = x^k$. Since T maps G into the abelian group $Z(G)$, we have $T(G') = 1$. That means that G' has a finite exponent (dividing k). Now, if $x \in Z(G)a_i$ and $y \in Z(G)a_j$, then $[x, y] = [a_i, a_j]$. Therefore G' is finitely generated. Since $|G'/G' \cap Z(G)| \leq \infty$, the subgroup $G' \cap Z(G)$ is also finitely generated, and being abelian and of finite exponent, it is finite. Thus G' is finite.   QED

Remark   It follows from the proof that there exists a bound for $|G'|$ depending only on k. However, the bound obtained by following the proof is much too big.

Proof of Theorem 3.7.   Let G have linear growth. Then it contains an infinite cyclic subgroup H of finite index, and replacing H by its normal core, we may assume that $H \lhd G$. Let $C = C_G(H)$. Then G/C is a group of automorphisms of H, and hence of order 1 or 2. By the previous theorem, C' is finite. Then C/C' is an abelian group with an infinite cyclic subgroup of finite index, therefore C/C' is an extension of a finite group by an infinite cyclic one, hence the same applies to C. Moreover, the finite normal subgroup of C obtained here is the set of all torsion elements of C, hence this subgroup, N say, is characteristic in C and normal in G. Finally, if $G = C$ then G/N is infinite cyclic, while if $|G : C| = 2$ then G/N is infinite dihedral.   QED

In the situation of Theorem 3.4 we can bound the index of a cyclic subgroup in terms of m. We saw that all long enough elements can be written as $us^k rv$. Now if w is long enough, we may, by taking "bites" out of t if necessary, assume that u and rv have length exactly m, and s^k has length at least m. Then s is an initial subword of a word of length m, and thus u and rv can be chosen in at most m ways each, and s can be chosen in at most m^2 ways. Moreover, given u, rv, s and

$n = l(w)$, the exponent k is uniquely determined, and w is determined with it. Thus $a_G(n) \le m^4$, and $s_G(n) \le m^4 n + A$, for some constant A (which accounts for the contribution of the short words). On the other hand, we saw that at least one s has infinite order. Since $l(s) \le m$, the subgroup $H := \langle s \rangle$ contains at least $2n/m - 1$ words of length n. Let $|G : H| = e$. The argument in the second paragraph of the proof of Proposition 2.5 shows that we can find representatives $\{a_i\}$ for the cosets of H of length at most e. Counting elements of the form $s^k a_i$ we see that $s_G(n) \ge e(2(n-e)/m - 1)$, and thus $e(2(n-e)/m - 1) \le m^4 n + A$. Letting n go to infinity, this implies that $e \le m^5/2$.

In [IS 87] it is shown that the best possible bound is m. However, the mere existence of a bound suffices for the main result of the next section.

3.2 Linear Growth Functions

In this section we show the following:

Proposition 3.11 *If G has linear growth, there are only finitely many groups with the same growth function as G.*

Proof Let H and G have the same growth function. If $a_G(m) > m$ for all m, then $s_G(m) > m(m+1)/2$, contradicting linearity. Thus for some m we have $a_G(m) \le m$ and we are in the situation of Theorem 3.4. This m is determined by the growth function of G, so now the preceding remarks show that H contains a cyclic subgroup C of bounded index. We may assume that $C \triangleleft G$, so what we have to show is that there are only finitely many groups which contain a normal infinite cyclic subgroup of bounded index, say $|H : C| \le k$. Then there are only finitely many possibilities for H/C, so it suffices to show that for each finite group R, and for an infinite cyclic C, there are only finitely many extensions H of C by R.

To that end, we review briefly the theory of group extensions. If $N \triangleleft G$, and $G/N \cong R$, we say that G is an *extension of N by R* (some authors prefer to say that G is an extension of R by N). Choose in G representatives a_r, $r \in R$, for the cosets of N. For convenience we choose $a_1 = 1$. Then each element of G has a unique expression as $x = a_r c$, $c \in N$. The mapping $x \to r$ is a homomorphism, and therefore $a_r a_s = a_{rs} c_{r,s}$, for some $c_{r,s} \in N$. Multiplication in G is given by

(M) $$a_r c \cdot a_s d = a_{rs} c_{r,s} c^{\alpha_s} d.$$

Here α_s is the automorphism that conjugation by a_s induces on N. Thus G is determined by the elements $c_{r,s}$, which are called a *factor set*, and by the automorphisms α_r. The choice $a_1 = 1$ implies $c_{r,1} = c_{1,s} = 1$. If we change the representatives a_r, say that the coset r is represented by an element $b_r = a_r c_r$, $c_r \in N$, $(c_1 = 1)$, then the automorphism α_r is changed by the inner automorphism that c_r induces, and we recommend to the reader to calculate the new factor set.

We now assume that N is abelian. Then all inner automorphisms are trivial, therefore the map $r \to \alpha_r$ is a homomorphism $\phi : R \to \mathrm{Aut}(N)$. This makes N an R-module. Calculating the product $(a_r a_s)a_t = a_r(a_s a_t)$ in the two indicated ways, we obtain

$$a_{rs}c_{r,s}a_t = a_{rs}a_t c_{r,s}^{\alpha_t} = a_{rst}c_{rs,t}c_{r,s}^{\alpha_t} = a_r a_{st}c_{s,t} = a_{rst}c_{r,st}c_{s,t},$$

and thus

(Ex) $$c_{rs,t}c_{r,s}^{\alpha_t} = c_{r,st}c_{s,t}.$$

Conversely, given a homomorphism $\phi : r \to \alpha_r$ of R to $\mathrm{Aut}(N)$, and a set of elements $c_{r,s}$ satisfying (Ex) and $c_{r,1} = c_{1,s} = 1$, we consider the set of formal products $a_r c$, with $r \in R$, $c \in N$ (here $\{a_r\}$ is a set of formal signs), and define multiplication by (M). It is easy to check that we get a group, and that the mapping $a_r c \to r$ is a homomorphism onto R, whose kernel is the subset $\{a_1 c\}$, so this subset is a normal subgroup, isomorphic to N by the map $a_1 c \to c$, with a factor group isomorphic to R. Given the homomorphism ϕ, there is always at least one factor set corresponding to it, namely $c_{r,s} = 1$. The corresponding extension is the semi-direct product of N by R determined by the given action of R on N. Moreover, given two factor sets $c_{r,s}$ and $d_{r,s}$ corresponding to the same ϕ, it is immediate that the product $c_{r,s}d_{r,s}$ is also such a factor set. It follows that the collection of all factor sets forms an abelian group; this group is denoted by $Z^2(R, N)$, and is known as the *group of 2-cocycles of R with coefficients in N*. Each factor set is termed a *2-cocycle*. The correspondence between extensions and 2-cocycles is not 1-1, because the same extension can define several factor sets. Changing the representatives a_r to $b_r = a_r c_r$, as above, changes the 2-cocycle $c_{r,s}$ to the cocycle $c_{r,s}c_r^{\alpha_s}c_s$. Now, given any map $r \to c_r$ of R to N (usually not a homomorphism), it is easy to check that $c_r^{\alpha_s}c_s$ is a 2-cocycle. Such a cocycle is termed a *2-coboundary*, and it is immediate that the set of coboundaries is a subgroup of $Z^2(R, N)$; this subgroup is denoted by $B^2(R, N)$. We see that changing representatives multiplies the corresponding cocycle by a coboundary, and conversely, two cocycles differing by a coboundary correspond to the same extension, with different choices of coset representatives. Thus each extension defines a unique element of the factor group $Z^2(R, N)/B^2(R, N)$. The latter group is denoted by $H^2(R, N)$, and is called the *second cohomology group of R with coefficients in N*.

At this stage we stop our development of extension theory; most group theory books contain chapters about it, which the interested reader can pursue, and in particular can understand where our terminology comes from. For our purposes it suffices to note that to show that there are only finitely many extensions of N by R, it suffices to find, first, that there are only finitely many homomorphisms $\phi : R \to \mathrm{Aut}(N)$, and, second, that for each ϕ the group $H^2(R, N)$ is finite.

In particular, to prove Proposition 3.11, we take $N = C$, an infinite cyclic group, and R any finite group. Since $\mathrm{Aut}(C) \cong C_2$, there are certainly only finitely many homomorphisms ϕ. We are not going to prove the finiteness of the cohomology groups, but just quote it from [Hu 67, Satz I.16.19(b)] (that theorem asserts that $H^2(R, N)$ is finite whenever R is finite and N is finitely generated as an R-module).   QED

4

The Growth of Nilpotent Groups

4.1 Polynomial Growth of Nilpotent Groups

Theorem 4.1 *Nilpotent-by-finite groups have polynomial growth.*

Proof By Proposition 2.5(c), we may assume that our group G is nilpotent. We employ induction on the Hirsch length $h(G)$ of G. If $h(G) = 1$, then G is finite-by-(infinite cyclic)-by-finite, and so its growth type is the same as of $\mathbb{Z}$, i.e. linear. Let G have a central series $1 = G_{r+1} \leq \cdots \leq G_1 = G$ with cyclic factors, and let $G_i = \langle G_{i+1}, x_i \rangle$, so that $G = \langle x_1, \ldots, x_r \rangle$. If G/G_2 is finite, then again it suffices to consider G_2. We thus may assume that G/G_2 is infinite, and then $h(G_2) = h(G) - 1$, and the induction hypothesis applies to G_2. Consider an element $x \in G$, written as a word of length n (or less) in the generators $\{x_i\}$, say $x = w_1 = y_{i_1} \cdots y_{i_n}$, where each y_i is either an x_j or an x_j^{-1}. We are going to rewrite x in the form $x = w_2 = x_1^e z$, for some integer e, where $z \in G_2$. We start by looking for the first occurrence of x_1 (or x_1^{-1}) that is to the right of another generator: say we have an occurrence of $x_2 x_1$, and we replace that by the equal product $x_1 x_2 [x_2, x_1]$. If x_2 was preceded by x_3, we now have the product $x_3 x_1$, which we replace by $x_1 x_3 [x_3, x_1]$. In each such replacement x_1 is pushed one position to the left, and we continue till it arrives at the extreme left, i.e. till the segment of w_1 that precedes it is itself a power of x_1 (possibly 1). We then look for the next occurrence of x_1; maybe it occurs in $x_4 x_1$, and we replace that in the same manner. Another possibility is that the next x_1 occurs right after the previous one, and in that case our first replacement caused the appearance of $[x_2, x_1] x_1$, which we replace by $x_1 [x_2, x_1][x_2, x_1, x_1]$. We continue this process till we have moved all occurrences of x_1 to the extreme left of our word, at which point we have our word w_2. Note that in this collection process we add only commutators; the number of occurrences of the original generators does not increase, though it may decrease, if we had some x_i appearing originally with both positive and negative exponents, because then we have cancellation at some stage.

It is clear that $|e| \leq n$, thus the number of possible values of the power x_1^e is not more than $2n + 1$. To bound the number of possible values of z, we want to know how many commutators we have there. First, commutators of the form $[x_i, x_1]$ can be created only when we move x_1 to the left of x_i, and such occurrence means that in w_1 we had x_1 to the right of x_i (not necessarily as an immediate right neighbour). The number of such pairs x_j, x_1 is at most n^2, so there are not more than n^2 commutators $[x_j, x_1]$. Suppose we already know that z contains not more than n^k commutators of weight k, and consider ones of weight $k + 1$. These have the form $[u, x_1]$, where u is a commutator of weight k, and such a commutator was created if at some stage u was created to the left of some x_1. The number of pairs u, x_1 is at most n^{k+1}, so by induction we get that for each weight t, there are not more than n^t commutators of weight t in z. Now if the nilpotency class of G is c, then all commutators of weight $c + 1$ or more are trivial, and we have to count only commutators of weights up to c, whose number is at most $n^2 + \cdots + n^c$. All these commutators lie in $G' \leq G_2$, and we can express them as words in the generators $\{x_2, \ldots, x_r\}$ of G_2. Let A be the maximum length of these words. The word z contains these commutators and those of the elements $\{x_2, \ldots, x_r\}$ that occur in the original element x. There are at most n of these latter occurrences, therefore the length of z is at most $A(n + n^2 + \cdots + n^c) < An^{c+1}$. Therefore $s_G(n) \leq (2n + 1)s_{G_2}(An^{c+1})$ is polynomial. QED

We can get from the above proof an explicit bound for the degree of polynomial growth of a nilpotent group, but it does not yield the best possible bound. To get that we have to do an elaborate bookkeeping while counting the number of added commutators. We first define the *rank* of a finitely generated abelian group as the number of infinite factors in its decomposition as a direct product of cyclic groups (thus the rank equals the Hirsch length). Then the result is

Theorem 4.2 *Let G be a finitely generated nilpotent group, let $r(i)$ be the rank of the abelian group $\gamma_i(G)/\gamma_{i+1}(G)$, and write $r = \sum i r(i)$. Then there exist constants C_1 and C_2, such that*

$$C_1 n^r \leq s_G(n) \leq C_2 n^r.$$

Proof Let G have class c. Since the torsion subgroup $\delta(G) = \delta_{c+1}(G)$ of G is finite, it follows from Proposition 2.5 and its proof, that it suffices to prove the theorem for $G/\delta(G)$. Moreover, for each i, the group $\delta_i(G)/\gamma_{i+1}(G)$ contains $\gamma_i(G)/\gamma_{i+1}(G)$ as a subgroup of finite index, and has $\delta_i(G)/\delta_{i+1}(G)$ as a factor group with a finite kernel. It follows that the abelian groups $\gamma_i(G)/\gamma_{i+1}(G)$ and $\delta_i(G)/\delta_{i+1}(G)$ have the same rank. Thus we may assume not only that G is torsion-free, but we will deal with the central series $\{\delta_i(G)\}$, which has torsion-free factors.

The derivation of the upper bound starts by choosing carefully a system of generators. Note that changing the generators can change the constants C_1 and C_2, but not the degree r. For each i, let the group $\delta_i(G)/\delta_{i+1}(G)$ be the direct sum of the cyclic groups generated by elements $x_{ij}\delta_{i+1}(G)$, where $j = 1,\ldots,r(i)$. The union of these sets x_{ij}, for $i = 1,\ldots,c = \mathrm{cl}(G)$, is our set of generators. We refer to x_{ij} as a generator of *weight* i. It is clear that each element of G can be written as a product $\Pi_{i=1}^c x_{i1}^{e_{i1}} \cdots x_{ir(i)}^{e_{ir(i)}}$, and we call this product the *canonical form* of that element. It is unique.

Let x be any element, written as a product of length at most n in the generators. We describe an algorithm which transforms x to the above canonical form. For convenience, let us write $x_1,\ldots,x_t$ for $x_{11},\ldots,x_{1r(1)}$. We start by collecting x_1 to the left, as in the previous proof, obtaining $x = x_1^{e_1}z$. Once this is done, we collect x_2 to the left of the word z, ending in a word $x_1^{e_1}x_2^{e_2}z_1$, then we collect x_3 to the left of z_1, and so on, till we have collected all the x_i, obtaining an expression $x = x_1^{e_1} \cdots x_t^{e_t}u$, where u is a word in x_{ij} for $i \geq 2$ and in various commutators. These commutators, in turn, can be expressed as words in the same x_{ij} (with $i \geq 2$), and we express them so. We end in an expression $x = ab$, where a looks like the part involving only the x_{1j} in the canonical form, and b is a word in the rest of the generators.

We now repeat the process, collecting to the left of b the generators x_{2j}. Then we collect the x_{3j}, etc. At each stage we add commutators which can be written as words involving only generators that were not collected yet, and in particular in the cth stage all such commutators are trivial, so all we do at that stage is rearrange the elements x_{cj}, and we have x in the canonical form.

Now for the bookkeeping part. As mentioned, each commutator $y = [y_1,\ldots,y_s]$, where each y_m is some x_{ij} (or its inverse), with $i = i(m)$ depending on m, can be written as a word in the x_{kj} with $k \geq \sum i(m)$. Since $s \leq c$, there are only finitely many such words. Let A be an upper bound for the lengths of all these words.

At the first stage of the collection commutators $[x_i, x_j]$ were formed whenever we had $i < j$ and x_i occurred to the right of x_j, and only in these circumstances. The number of such pairs i, j is at most n^2, and this is an upper bound for the number of the commutators (for simplicity, we ignored the possibility that we have not x_i, say, but its inverse; this does not affect the bookkeeping). After such a commutator was formed, some x_k may have been collected past it, creating a commutator $[x_i, x_j, x_k]$. The number of occurrences of these commutators is at most n^3, and generally we obtain that the number of commutators of weight s in the word u is at most n^s. Replacing them by their expressions in terms of the generators, we see that the total number of occurrences of generators of weight s in the word b is not more than $A(n + n^2 + \cdots + n^s) \le cAn^s$ (the term n is added to take care of those generators of weight s that were present in our word to begin with).

Suppose that, after we have collected all generators of weight less than k, we have in the uncollected part, for each s, at most Bn^s generators of weight s, for some constant B. Then in the next stage we collect the generators of weight k; let these be $y_1, \ldots, y_t$ (here of course $t = r(k)$, not $r(1)$). The number of times that these generators occur does not change during this collecting. The process will first yield many commutators of type $[x_{ij}, y_{i_1}, \ldots, y_{i_s}]$, where $i \ge k$, and $s < c$. Since x_{ij} occurs at most Bn^i times and for each y_{i_l} there are at most Bn^k possibilities, the number of occurrences of commutators of the this type, for a fixed i and s, is at most $B^{s+1}n^{i+sk}$. Each such commutator lies in $\delta_w(G)$, where $w = i + sk$. Writing these commutators in terms of the generators x_{ij}, for $i \ge w$, and allowing for the various values of i and s, we see that each generator of weight w occurs at most $c^2AB^wn^w$ times, and adding its previous number of occurrences, which is Bn^w or less, we that the constant B is changed to $B + c^2AB^c$, and with this new constant we proceed to the next stage.

When the collection is finished and we have x in its canonical form, then for each i the number of the generators of weight i in this form is at most Dn^i, where the constant D depends on G, on our choice of generators, and on the way that we express the commutators of the generators, but D does not depend on x. If the part of weight i in the canonical form is $x_{i1}^{e_{i1}} \cdots x_{ir(i)}^{e_{ir(i)}}$, that means that $\sum |e_{ij}| \le Dn^i$. Then the number of values of this weight i part is the same as in the abelian group $\delta_i(G)/\delta_{i+1}(G)$. The latter we already know to be some polynomial f_i of degree $r(i)$, so this number of values is at most $f_i(Dn^i)$, and the number of possible elements x is at most $\Pi_{i=1}^{c} f_i(Dn^i) \le C_2n^r$, as claimed.

For deriving the lower bound we consider an arbitrary finite set X of generators of G. We apply induction on $c = \mathrm{cl}(G)$, the result being true for $c = 1$. For $c > 1$, we first prove, also by induction, the following claim

Lemma 4.3 *There exists a constant A, and a set Z of generators of $\gamma_c(G)$, such that for each $z \in Z$ and for each positive integer n, we can write z^n as a word of length at most $An^{1/c}$.*

Proof The lemma is obvious for $c = 1$ (with $A = 1$). Let $c > 1$. Write $N = \gamma_c(G)$. Suppose that we have found already a constant B and a set Y, such that the images of Y in G/N generate $\gamma_{c-1}(G)/N$ and fulfill our requirement in G/N. By Lemma 2.14, N is generated by the conjugates of commutators of the form $[x, y]$, $x \in X$, $y \in Y$. But these commutators lie in $N \leq Z(G)$, so are equal to their conjugates, and these commutators themselves are generators of N.

Let $z = [x, y]$ be one of these generators. Let m satisfy $2n^{1/c} \geq m > n^{1/c}$, and write $n = qm^{c-1} + s$, with $q \geq 0$ and $0 \leq s < m^{c-1}$. Then $0 \leq q < m$. By induction, applied to $G/\gamma_c(G)$, there exist two elements u, v of lengths $\leq Bm$ such that $y^{m^{c-1}} = ut$, $y^s = vw$, $t, w \in N$. Then $[x, y]^n = [x, y^n] = [x, y^{qm^{c-1}+s}] = [x^q, y^{m^{c-1}}][x, y^s] = [x^q, ut][x, vw] = [x^q, u][x, v]$ has length at most $4Bm + 2q + 2 \leq (4B+4)m \leq (8B+8)n^{1/c}$.   QED

Now return to the main proof. Below, D, E, F stand for constants independent of n. Among the elements of Z there are $r(c)$ independent ones, generating a free abelian subgroup of rank $r(c)$. Applying the lemma, the powers z^i, $0 < i \leq n^c$ of each generator produce n^c elements of length not more than An, and products of these powers yield at least $n^{cr(c)}$ elements of length at most $r(c)An$. The induction hypothesis supplies us with at least $Dn^{r-cr(c)}$ elements of length n which are all different mod N. The products of these two sets of elements yield Dn^r elements of length at most En. This ends the proof.   QED

Corollary 4.4 *Let G be a finitely generated nilpotent group of growth degree d. Then G contains a finite index subgroup H such that $\mathrm{cl}(H) < \sqrt{2d}$ and H can be generated by d elements.*

Proof By Theorem 2.12, G contains a finite index torsion-free subgroup H, and G and H have the same growth degree. Consider the central series $\{\delta_i(H)\}$, and recall that $\delta_i(H)/\delta_{i+1}(H)$ and $\gamma_i(H)/\gamma_{i+1}(H)$ have the same rank $r(i)$. Lemma 4.9 below shows that $r(1) \geq 2$, unless $d = 1$. Let $\mathrm{cl}(H) = c$, then Theorem 4.2 implies $d = \sum ir(i) \geq 1 + \sum i = c(c+1)/2 + 1$. Thus $c < \sqrt{2d}$. Moreover, the factor group $\delta_i(H)/\delta_{i+1}(H)$

is generated by r_i elements, and thus H can be generated by $\sum r_i \le d$ elements.   QED

If H is nilpotent and K is finite, then H and $H \times K$ have the same degree, therefore the index of a nilpotent subgroup cannot be bounded in terms of the degree alone. However, if we have $s_G(n) \le Cn^d$, then the index of a nilpotent subgroup of G can be bounded by a function of d and C. Moreover, if G is torsion-free, then the same index can be bounded in terms just of d. These results will be proved in Chapter 9 below.

P. Pansu [Pa 83] proved a result sharper than Theorem 4.2: *if G is a finitely generated nilpotent (or nilpotent-by-finite) group, and r is as above, there exists a constant A, such that $s_n(G)$ is asymptotic to An^r,* i.e. $\frac{s_n(G)}{An^r} \to 1$ *as $n \to \infty$*. If $\mathrm{cl}(G) = 2$, then M. Stoll [St 98] has refined this by showing that there exists a further constant B such that $|s_n(G) - An^r| \le Bn^{r-1}$. On the other hand, in [St 96] Stoll shows that even for class 2 nilpotent groups the growth function can be of very irregular behaviour. The proofs of these results will not be given here. It is easy to deduce from Pansu's result that if G is nilpotent-by-finite, then $a_n = o(s_n)$. R. Tessera [Te 07] has shown that a sharper result can be deduced already from Theorem 4.2.

Lemma 4.5 *Let G be nilpotent-by-finite. Then for each positive number α there exists a constant $C = C(\alpha)$ such that $s_G(\alpha n) \le Cs_G(n)$.*

In the notation of Theorem 4.2, we have, for all n: $s(\alpha n) \le C_2(\alpha n)^r \le (C_2/C_1)\alpha^r C_1 n^r \le (C_2/C_1)\alpha^r s(n)$.

Notation   For each $x \in G$, $B(x, \rho)$ is the ball of radius ρ around x, i.e. the set of elements such that $l(y^{-1}x) \le \rho$, and $S(m, n) = \{x \in G \mid m < l(x) \le n\}$.

Note that $B(x, \rho)$ contains exactly $s(\rho)$ elements of G.

Lemma 4.6 *Let G be nilpotent-by-finite. There exists a constant $c > 0$ such that for all n and all k we have $|S(n - k, n)| \ge c|S(n, n + k)|$.*

Proof   Let us put inside $S(n - k, n)$ as many disjoint balls of radius $k/2$ as we can, say t. Then each point of $S(n - k, n)$ is at distance at most k from the centre of one of these balls. Each element of $S(n, n + k)$ is at distance k from some element of $S(n - k, n)$, and therefore $S(n, n + k)$ is covered by the t balls of radius $2k$ around the same centres as the balls in $S(n - k, n)$. Thus, using the notation of the previous lemma: $|S(n, n + k)| \le ts(2k) \le tC(4)s(k/2) \le C(4)|S(n - k, n)|$.   QED

Theorem 4.7 *Let G be nilpotent-by-finite, and let r be its growth degree, as in Theorem 4.2. Then there exists a number $r - 1 \leq s < r$ such that $a_G(n) \leq An^s$, for some constant A and all n.*

Proof Let $2^i \leq n$. The previous lemma shows that $|S(n-2^i, n-2^{i-1})| \geq c|S(n - 2^{i-1}, n)|$, and therefore $|S(n - 2^i, n)| \geq (1 + c)|S(n - 2^{i-1}, n)$. Iterating this yields $|S(n - 2^i, n| \geq (1+c)^i|S(n - 1, n)| = (1+c)^i a_G(n)$. Taking $i = \lfloor \log_2 n \rfloor$, we obtain $C_2 n^r \geq s_G(n) \geq |S(n - 2^i, n)| \geq (1 + c)^i a(n) \geq (1 + c)^{\log_2 n - 1} a(n) = (1/(1 + c))(n)^{\log_2(1+c)} a(n)$.

Thus $a(n) \leq (1 + c)C_2 n^{r - \log_2(1+c)}$. QED

The inequality $r - 1 \leq s$ is obvious, otherwise, by summing, we contradict the lower bound for $s(n)$ in Theorem 4.2.

If we try to evaluate s explicitly from the above proof, we find that it may be very close to r. On the other hand, for abelian groups both $s(n)$ and $a(n)$ are polynomials, of degrees r and $r - 1$ respectively. For groups of class 2 this need not be the case, but Stoll's result for this case shows that we can take $s = r - 1$.

Problem Give good bounds for s above. In particular, can we always have $s = r - 1$?

4.2 Groups of Small Degree

In the previous chapter we saw that all groups of linear growth lie in the same commensurability class. For nilpotent groups we can now establish similar properties for several bigger growth types.

Proposition 4.8 *Within the class of nilpotent-by-finite groups we have:*

(a) *All groups of quadratic growth lie in the same commensurability class, and all groups of cubic growth lie in the same commensurability class.*

(b) *All groups of growth of degree 4 lie in two commensurability classes, and all groups of degree 5 lie in two commensurability classes.*

(c) *The groups of degree 6 lie in three commensurability classes.*

(d) *The groups of degree 7 lie in five commensurability classes.*

(e) *There are infinitely many commensurability classes of groups of degree 8.*

The proof applies the following simple lemma.

Lemma 4.9 *In a non-abelian group G, the factor group $G/Z(G)$ is not cyclic.*

Proof Assume that $G/Z(G)$ is generated by the coset $xZ(G)$. Then G is generated by $Z(G)$ and x. Since the generators commute, G is abelian.   QED

The lemma implies that for any non-abelian nilpotent group G, its *abelianization*, i.e. the factor group G/G', is not cyclic, otherwise $G/\gamma_3(G)$ would violate the lemma.

Proof of 4.8.   By Theorem 2.12, any nilpotent-by-finite group contains a torsion-free nilpotent group of finite index. Thus we assume that G is torsion-free nilpotent.

(a) If G has degree $r = 2$, then Theorem 4.2 shows that G is abelian and isomorphic to $\mathbb{Z} \times \mathbb{Z}$. Let $r = 3$. Theorem 4.2 shows that G is either free abelian of rank 3, or $\gamma_2(G)$ is infinite cyclic central, with factor group of rank 1. But in the latter case $\delta_2(G)$ is central (by Proposition 2.22), with an infinite cyclic factor group, contradicting the lemma.

(b) If $r = 4$, 4.2 shows that G is either free abelian of rank 4, or G/G' has rank 2 and G' is infinite cyclic central. In the latter case, since $G' \leq Z(G)$, the group $G/Z(G)$ is abelian of rank 0, 1, or 2. Rank 0 means that $G/Z(G)$ is finite, G and $Z(G)$ are commensurable, and since $r = 4$, $Z(G)$ is free abelian of rank 4. If the rank is 1, then $G/Z(G)$ contains an infinite cyclic subgroup $H/Z(G)$ of finite index. Then H is abelian, by Lemma 4.9, and commensurable with G, and again it is isomorphic to $\mathbb{Z}^4$. If the rank is 2, then, since G has Hirsch length 3, $Z(G)$ is infinite cyclic. Write $G/Z(G) = F \times T$, where F is free abelian of rank 2 and T is finite. Let $F = H/Z(G)$, then H is of finite index, and there exist $x, y \in G$ such that $H = \langle x, y, Z(G) \rangle$. Write $K = \langle x, y \rangle$. Then $H' = K' = \langle [x, y] \rangle$. We may assume that H is not abelian, so that H' is a non-trivial subgroup of $Z(G)$, hence $|Z(G) : H'|$ is finite. But $H = KZ(G)$, therefore $|H : K| = |Z(G) : Z(G) \cap K| \leq |Z(G) : K'|$ is finite, and $|G : K|$ is finite. Thus G is commensurable with K, and we conclude the proof of this case by identifying K.

Proposition 4.10 *A two-generator group with a central infinite cyclic commutator subgroup is isomorphic to the free nilpotent group of class 2 and rank 2.*

Proof Let K be the group in question, say $K = \langle x, y \rangle$, and let F be the free nilpotent group of class 2 and rank 2. If K/K' is cyclic, then K

is abelian (by Lemma 4.9), while we are assuming that $K' \neq 1$. And if K/K' contains an element of finite order, then, being an abelian group with two generators, K/K' is the direct sum of two cyclic groups, at least one of which is finite. Lifting the generators of these cyclic groups to K, we see that K is generated by two elements, say u, v, one of which is of finite order mod K', and therefore modulo the centre. This implies that $[u, v]$, which generates K', is of finite order, again a contradiction.

Thus K/K' is free abelian of rank 2, and the Hirsch length of K is 3. Exactly the same arguments show that $K/Z(K)$ is free abelian of rank 2, and since $h(K) = 3$, this implies that $Z(K)$ is also infinite cyclic, say $Z(K) = \langle z \rangle$. If the inclusion $K' \leq Z(K)$ is proper, then z has finite order mod K'. Thus $Z(K) = K'$, we may assume that $z = [x, y]$, and each element of K has a unique expression as $x^i y^j z^k$. This identifies K with F.   QED

Abstractly, F is given by the presentation $F = \langle x, y \mid [x, y, x] = [x, y, y] = 1 \rangle$. Concretely, it is represented as the group $U(3, \mathbb{Z})$, where we recall from Section 2.4 that this is the group of upper unitriangular three-by-three integer matrices (i.e. the integral matrices with 1 on the main diagonal and 0 below it). This group is often called the *discrete Heisenberg group* (the *Heisenberg group* is $U(3, \mathbb{R})$, the three-by-three unitriangular matrices over the reals).

The continuation of the proof of 4.8 requires some familiarity with the theory of alternating (skew-symmetric) bilinear forms (some of the arguments involving alternating forms can be replaced by direct commutator calculations).

Let $r = 5$. As before, we see that G is either free abelian of rank 5, or G/G' has rank 3, and G' is infinite cyclic and central. We assume that G is of the latter type. Then $N := \delta_2(G)$ is infinite cyclic and $K := G/N$ is free abelian of rank 3. Commutation defines an alternate bilinear form from K into N, and this form determines G completely. We can embed N and K in one-dimensional and 3-dimensional $\mathbb{Q}$-vector spaces, say V and W, and extend the bilinear form to one from W to V. Since V is one-dimensional, we can regard the form as one from W to $\mathbb{Q}$, and since W has odd dimension, the form is degenerate, i.e. there exists $0 \neq w \in W$ which is orthogonal to all of W. Some integer multiple of w lies in K, and translating this back to G means that G contains a central element outside N. Then both $Z(G)$ and $G/Z(G)$ are free abelian of rank 2. If the latter is generated by $xZ(G)$ and $yZ(G)$, say, then G' is generated by $[x, y]$. We can choose generators z, w for $Z(G)$ such that $[x, y] = w^n$, for some n. Then $G = \langle z \rangle \times \langle x, y, w \rangle$, and has the subgroup $\langle z \rangle \times \langle x, y \rangle$, of index n. Here x and y generate a free nilpotent group of class and rank 2, and G is commensurable with the direct product of that group and an infinite cyclic one.

(c) If $r = 6$ then either G is free abelian of rank 6, or G' is central, and either infinite cyclic, with G/G' of rank 4, or both G' and G/G' have rank 2. But in the latter case $G/Z(G)$ is generated by two elements, say $xZ(G)$ and $yZ(G)$, and then G' is generated by $[x, y]$ and is not of rank 2. Thus we may assume that G' is cyclic and G/G' has rank 4. Replacing G' by $\delta_2(G)$, we see that G is an extension of a central cyclic subgroup N by a free abelian group K of rank 4. As above, we embed N and K in vector spaces V and W, of dimensions 1 and 4, and commutation in G induces an alternating map from W to V. There are two possibilities.

First, the radical of the form may have dimension 2. That means that $Z(G)$ is free abelian of rank 3, and $G/Z(G)$ is free abelian of rank 2. Then we can find generators x, y, z, u, v for G such that x and y generate $G \mod Z(G)$, and the other three elements generate $Z(G)$. Moreover, these generators can be chosen to satisfy $[x, y] = z^n$, for some n, and then G contains the finite index subgroup $\langle u, v \rangle \times \langle x, y \rangle$, with the first direct factor being free abelian of rank 2 and the second isomorphic to the discrete Heisenberg group.

The second possibility is that the alternating form on W is non-degenerate. Then we can choose generators as above such that z generates $Z(G)$ and the other four generate $G \mod Z(G)$, and then we have $[x, y] = z^n$, $[u, v] = z^k$, for some n and k, and G contains the finite index subgroup $\langle x^k, y, u^n, v \rangle$, which is isomorphic to the group with 4 generators a, b, c, d with relations $[a, b] = [c, d]$ and all other commutators are 1. That group is of class 2 (why?). It can be described more concretely as follows: start with two copies of the Heisenberg group, generated by $\{a, b\}$ and by $\{c, d\}$ respectively, form their direct product, and factor out that product by the central subgroup generated by $[a, b][c, d]^{-1}$. We describe that construction by saying that G is the central product of the two copies of the Heisenberg group, identifying $[a, b]$ and $[c, d]$.

(d) Let $r = 7$. As above, we see that the possibilities for G are: free abelian of rank 7; or of class 2, and then $\delta_2(G)$ is central, and either cyclic or free abelian of rank 2, with a factor group free abelian of rank 5 or 3, respectively; or of class 3, with $\delta_2(G)/\delta_3(G)$ and $\delta_3(G)$ infinite cyclic, and $G/\delta_2(G)$ free abelian of rank 2.

When G is of class 2, we define, as above, vector spaces W and V containing $G/\delta_2(G)$ and $\delta_2(G)$, respectively. The dimensions of these spaces are either 5 and 1, or 3 and 2, and there is an alternating form from W to V.

If $\dim(V) = 1$, the radical of the form is of dimension 3 or 1. In the first case $Z(G)$, which is always free abelian, has rank 4, and we can choose generators x, y, z, u, v, w for G, such that $Z(G) = \langle z, u, v, w \rangle$, and $[x, y] = z^n$, for some $n \neq 0$. Then x and y generate a Heisenberg subgroup H, and the direct product of H with the free abelian subgroup $\langle u, v \rangle$ has a finite index in G. If the radical is 1-dimensional, then, similarly to case **c.**, we can find a subgroup of finite index which is the direct product of an infinite cyclic group by a central product of two Heisenberg groups.

Let $\dim(V) = 2$, then we can write $G' = \langle u, v \rangle$, and then for any two elements x, y of G we write $[x, y] = u^n v^m$. The maps $(x, y) \to n$ and $(x, y) \to m$ induce a pair of alternating forms on W. Since $\dim(W) = 3$, these forms are degenerate. If some $0 \neq w \in W$ belongs to the radical of both forms, that means that $G/Z(G)$ has rank 2, generated by $xZ(G)$ and $yZ(G)$, say. But then $G' = \langle [x, y] \rangle$, contradicting G' having rank 2. Therefore we can find a basis x, y, z for W, such that x and y span the radicals of the two forms, respectively. Abusing notation a little, we replace these vectors first by integer multiples lying in $G/\delta_2(G)$, and then by representatives from G, and we find that $\langle x, y, z \rangle$ is a finite-index subgroup defined by the relations $[x, z] = u$, $[y, z] = v$, and all other commutators are 1.

Finally, let G be of class 3. We choose generators x, y, z, w such that the first two generate G ($\mod \delta_2(G)$), and the other two generate $\delta_2(G)/\delta_3(G)$ and $\delta_3(G)$, respectively. We may also assume that $[x, y] = z^n$, for some n, and $[x, z] = w^k$, $[y, z] = w^l$, with $k \neq 0$. Then $u := x^{-l} y^k \in Z(G)$, and $\langle x, x^{-l} y^k \rangle$ is of finite index, and isomorphic to the group $\langle a, b, c, d \mid [a, b] = c, [a, c] = d \rangle$, and all other commutators are 1. This ends case **(d)**.

(e) Here we are not going to present detailed proofs, but rather quote some results from [GSS 82]. In that paper, torsion-free groups of class 2 in which G' has rank 2 and G/G' is free abelian of rank 4 are classified. These groups have degree 8. With each such group a binary integral quadratic form $aX^2 + bXY + cY^2$ is associated. Recall that the *discriminant* of the form is $b^2 - 4ac$. All such forms can occur, and two groups are commensurable if either the two associated forms are both 0, or neither is 0, and the ratio between the discriminants is a non-zero rational square. Since any integer which is congruent to 0 or 1 ($\mod 4$) is a discriminant, and Q^*/Q^{*2} is infinite, our claim follows.   QED

In the notation of Theorem 4.2, the Hirsch length of G is $\sum r(i)$. It is clear from the last proof that classifying nilpotent-by-finite groups of low growth up to commensurability is the same as classifying torsion-free nilpotent groups of small Hirsch length. For more on that problem, the reader may consult [GS 84] and [Se 83, Ch. 11.D].

5

The Growth of Soluble Groups

5.1 Soluble Groups of Polynomial Growth

As already mentioned, one of the major results of our subject is M. Gromov's converse of Theorem 4.1: *a group of polynomial growth is nilpotent-by-finite* [Gro 81]. This was proved first by J. Milnor [Mi 68] and J. Wolf [Wo 68] for soluble groups, and that result is our next goal. We start with:

Theorem 5.1 *Let G be a finitely generated group of subexponential growth. Then the commutator subgroup G' of G is also finitely generated.*

Corollary 5.2 *A finitely generated soluble group of subexponential growth is polycyclic.*

Proof Let G be soluble of polynomial growth. The preceding theorem and induction imply that all commutator subgroups of G are finitely generated, and then so are the abelian factor groups $G^{(i)}/G^{(i+1)}$. Thus these factor groups are polycyclic, and since only finitely many of them are non-trivial, so is G. QED

Proof of Theorem 5.1. Since G/G' is a finitely generated abelian group, it is a direct product of finitely many cyclic groups. It will thus suffice to show that if $N \triangleleft G$ and G/N is cyclic, then N is finitely generated. Since all finite index subgroups of G are finitely generated, we may assume that G/N is infinite cyclic. Let xN generate G/N. Given any generators $\{x_1, \ldots, x_d\}$ of G, we can write them in the form $x_i = x^{e_i} y_i$, where $y_i \in N$, and then $x, y_1, \ldots, y_d$ generate G. Then N contains the normal closure, K say, of the $y's$. But G/K is generated by the images of the generators of G, hence by xK, so G/K is infinite cyclic, and G/N is an infinite cyclic factor group of it, which is possible only if $K = N$.

Let K_i be the subgroup generated by all the conjugates $x^{-n} y_i x^n$, then $N \geq \langle K_1, \ldots, K_d \rangle$, and the latter subgroup contains $y_1, \ldots, y_d$ and is invariant under conjugation by all the generators of G, hence it is equal to N. It will thus suffice to prove that each K_i is finitely generated. To that end, we write y for y_i, and consider the products $xy^{e_1} x y^{e_2} x \cdots y^{e_n}$, where each e_i is 0 or 1. There are 2^n such words, all of length $2n$ or less, and the subexponentiality implies that if n is large enough, two of these words are equal. Let us consider the minimal n at which equality occurs, say, the word above equals a similar one with exponents f_i. By minimality, $e_n \neq f_n$. Write $y(k) = x^k y x^{-k}$, and write the equality in the form $y(1)^{e_1} y(2)^{e_2} \cdots y(n)^{e_n} x^n = y(1)^{f_1} \cdots y(n)^{f_n} x^n$. Since $e_n \neq f_n$, this shows that $y(n)$ can be expressed as a product of $y(1), \ldots, y(n-1)$. Therefore $y(n+1) = xy(n)x^{-1}$ can be expressed in terms of $y(2), \ldots, y(n)$, and substituting the expression of $y(n)$ we see that $y(n+1)$ also belongs to the subgroup generated by $y(1), \ldots, y(n-1)$, and an obvious induction shows that all the $y(n)$, for $n > 0$, belong to the same subgroup. Replacing x by its inverse, we see that the subgroup generated by the $y(n)$ for negative n is also finitely generated, hence so is K_i.   QED

The above result and argument are due to Milnor. For polycyclic groups Wolf proved the following:

Theorem 5.3 *A polycyclic group of subexponential growth is nilpotent-by-finite.*

Together with Theorem 4.1 this implies:

Corollary 5.4 *Let G be a finitely generated soluble group. Then the growth of G is either exponential or polynomial, and the latter occurs if and only if G is nilpotent-by-finite.*

Proof of Theorem 5.3. Wolf's proof applies the theory of Lie groups. The following proof is due to H. Bass [Bs 72], and may be said to be purely algebraic, if we include under that term a few elementary facts about algebraic numbers. Let G be polycyclic of subexponential growth. Going down a normal series of G with cyclic factors, we arrive at a subgroup of finite index which has an infinite cyclic factor group, and we may replace G by this subgroup, and thus assume that there exists a subgroup $H \lhd G$, such that G/H is infinite cyclic, generated by xH, say.

Then H has a smaller Hirsch length than G, so by induction we may assume that H contains a finite index nilpotent subgroup N. Since H is finitely generated, N contains a characteristic subgroup of H which is still of finite index, and by replacing N by this subgroup, we may assume that $N \triangleleft G$. Then $G/N = H/N \cdot \langle xN \rangle$, so $|G : N\langle x \rangle| \leq |H/N|$ is finite, and we replace G by $\langle N, x \rangle$. Thus G is an extension of the finitely generated nilpotent group N by the infinite cyclic group generated by x. We are going to find a central series N_i of N, consisting of normal subgroups of G, and an integer k such that x^k centralizes each factor N_{i+1}/N_i, and then the series N_i, supplemented by the link $N \triangleleft K$, is a central series for $K := \langle N, x^k \rangle$, so that K is nilpotent, and $|G : K| = k$, concluding our proof.

To find the series N_i, we first note that there exist central series of N, such as the lower or upper central series, whose terms are normal in G. In each central series of N the number of infinite factors does not exceed the Hirsch length of N, therefore we can find a central series, say L_i, consisting of normal subgroups of G, with a maximal number of infinite factors. Each factor L_{i+1}/L_i is a finitely generated abelian group, with a torsion subgroup T_i/L_i, say, and we refine our series further by inserting the subgroups T_i.

The resulting series, which we call N_i, is such that if N_{i+1}/N_i is infinite, and if $M \triangleleft G$ and $N_i \leq M \leq N_{i+1}$, then either $M = N_i$ or N_{i+1}/M is finite. We now want to find the aforementioned integer k. Obviously, it suffices to find for each i an integer k_i such that x^{k_i} centralizes N_{i+1}/N_i, because then we can take k as the product of them. Also obviously such a k_i exists if the relevant factor is finite, so we assume that N_{i+1}/N_i is infinite. By our construction, this factor group is free abelian of finite rank, and it is convenient to write it additively, and to regard it as a $\mathbb{Z}[x]$ module, say M. We extend M to a $\mathbb{Q}[x]$ module V by tensoring with $\mathbb{Q}$. In less sophisticated language this means that we consider a basis for M, and let V be the vector space over $\mathbb{Q}$ with the same basis, and the same action of x on this basis. We regard M as a subgroup of V. Each element of V is a linear combination of the basis elements, and multiplying by a common denominator of the rational numbers that are the coefficients in this expression, we see that some integer multiple of this element lies in M, and thus V/M is a torsion group.

We claim that V is an irreducible $\mathbb{Q}[x]$ module. Indeed, let W be a non-trivial invariant subspace. It follows from the previous paragraph that $W \cap M \neq 0$. Then $M/W \cap M$ is finite, therefore $V/W \cap M$ is torsion, and so is V/W, but that is possible only if $W = V$.

We now recall the famous

Schur's Lemma *The endomorphism ring of an irreducible module V is a division ring.*

Proof Let $\alpha \neq 0$ be any endomorphism. Then $\mathrm{Ker}(\alpha) \neq V$, and $\mathrm{Im}\,(\alpha) \neq 0$: therefore $\mathrm{Ker}(\alpha) = 0$ and $\mathrm{Im}\,(\alpha) = V$. Thus α is an invertible map, and it is immediate that the inverse is also an endomorphism, so we are done. QED

Let α be the linear transformation that x induces on V. By the lemma, α lies in a division ring, and since V is finite dimensional, so is this ring. Therefore α generates some finite field extension of $\mathbb{Q}$, say F, and we may embed F as a subfield of the complex field $\mathbb{C}$. Since $F/\mathbb{Q}$ is a finite extension, α corresponds in this embedding to some algebraic number. Moreover, the various algebraic conjugates of this number are obtained by considering all embeddings of F into $\mathbb{C}$. Since α acts on M, it maps the elements of a basis of M to integer linear combinations of themselves, and thus it is represented by an integral matrix. Then α satisfies the characteristic polynomial of this matrix, so that it corresponds to an algebraic integer.

Lemma 5.5 *If all the conjugates of an algebraic integer τ have absolute value 1 at most, then τ is a root of unity.*

Proof Let σ be any power of τ, then all conjugates of σ also have absolute value at most 1. If τ has degree n, then $|\mathbb{Q}(\tau) : \mathbb{Q}| = n$, and $\sigma \in \mathbb{Q}(\tau)$, so σ also has a degree n or less. The minimal polynomial of σ is of degree at most n, and a typical coefficient of it is, apart from sign, the kth symmetric function of the conjugates of σ: therefore it has absolute value at most $\binom{n}{k}$. It follows that there are only finitely many polynomials that can occur as the minimal polynomial of σ, and therefore there are only finitely many powers of τ, which means that τ is a root of unity. QED

Return to the proof of the theorem. If the algebraic integer corresponding to α in the embedding above is a root of unity, that means that some power of the transformation α is the identity, i.e. some power of x is trivial on M, which is what we need. So, assuming that α does not correspond to a root of unity, the lemma tells us that we can find some embedding under which $|\alpha| \neq 1$. Replacing α by its inverse, if necessary, we may assume that $|\alpha| > 1$. Passing to a power of α (and of x) we may assume that $|\alpha| > 2$.

Recall the proof of Theorem 5.1. Arguing as there, with our element x and taking $y \in N_{i+1}$, $y \notin N_i$, we get that for some $0 \neq v \in V$ we have $v(\alpha^n) = v(e(1)\alpha + \cdots + e(n-1)\alpha^{n-1})$, with $|e(i)| \leq 1$. Since all non-zero endomorphisms of V are 1-1, this means that $\alpha^n = e(1)\alpha + \cdots + e(n-1)\alpha^{n-1}$. But $|e(1)\alpha + \cdots + e(n-1)\alpha^{n-1}| \leq |\alpha| + \cdots + |\alpha|^{n-1}| = \frac{|\alpha|^n - 1}{|\alpha| - 1} < |\alpha|^n$, a contradiction. The proof of Theorem 5.3 is done.   QED

5.2 Uniform Exponential Growth of Soluble Groups

D.V. Osin has further refined Corollary 5.4 by showing that soluble groups of exponential growth have uniform exponential growth [Os 03]. An alternative proof of that is given in [Br 07]. The important special case of polycyclic groups is discussed in [Al 02]. Here we will indicate only the reduction of the proof to the polycyclic case. On the way we will encounter an interesting phenomenon: for large classes of groups, not only are all the groups in these classes of uniform exponential growth, but the growth rate $\Omega(G) \geq C > 1$, where the constant C depends only on the class, not on G. This is sometimes described by saying that the relevant class has *uniformly uniform exponential growth*.

The first step in that is the following result, which is essentially a slight modification of the proof of Theorem 5.1.

Proposition 5.6 *Let G be a finitely generated abelian-by-cyclic group. If $\Omega(G) < \sqrt[5]{2}$, then G is polycyclic.*

Proof Choose a set of generators S such that $\omega(G, S) < \sqrt[5]{2}$. Let A be a normal abelian subgroup with a cyclic factor group. One of the elements of S, say x, lies outside A. The commutator subgroup G' is contained in A, and since G/G' is a finitely generated abelian group, it suffices to show that the abelian subgroup G' is finitely generated. Here G' is the normal closure of the finite set $\{[s, t] \mid s, t \in S\}$. Let y be one of the commutators $[s, t]$. If for all n all products $x y^{e_1} x y^{e_2} x \cdots y^{e_n}$, for $e_i = 0, 1$, differ from each other, then $s(5n) \geq 2^n$, and $\omega(G, S) \geq \sqrt[5]{2}$, a contradiction. Thus for some n there is an equality $x y^{e_1} x y^{e_2} x \cdots y^{e_n} = x y^{f_1} x y^{f_2} x \cdots y^{f_n}$, with $e_n \neq f_n$. The argument of Theorem 5.1 shows that the normal closure K_y of y in $\langle x, y \rangle$ is finitely generated, and since $y \in A$ and A is abelian, K is also the normal closure in $\langle A, x \rangle$. Let uA be a generator of G/A. Then $x = u^e z$, for some integer $e \neq 0$ and some $z \in A$. We may assume that $e > 0$. Then $\langle y \rangle^G = K K^u \cdots K^{u^{e-1}}$ is also finitely generated. Since G' is the product of all the subgroups $\langle y \rangle^G$, it is finitely generated.   QED

Lest the readers think that our result is empty, let us reassure them that there are many abelian-by-cyclic groups that are not polycyclic, e.g. wreath products $A \wr \mathbb{Z}$, where A is any non-identity abelian group. According to the calculations in [Jo 91], for $G = C_2 \wr \mathbb{Z}$, the wreath product of a group of order 2 by $\mathbb{Z}$, with the natural two generators, we have $\omega(G, S) = \frac{1+\sqrt{5}}{2} = 1.618\cdots$ (the golden ratio) (see also Setion 14.2 below), while $\sqrt[5]{2} = 1.149\cdots$. The best possible value of the bound in Proposition 5.6 is somewhere between these values.

Definition Let G be generated by S. A subgroup H of G has S-*depth* n, if H can be generated by elements of S-length at most n.

Lemma 5.7 *In the situation and notation of the definition, let G and H be finitely generated, then $\omega(G, S) \geq \Omega(H)^{1/n}$.*

This is obvious.

In the same situation, we let S_i denote the set of all simple commutators of weight i in the elements of S. By Lemma 2.16, this is a set of normal generators for $\gamma_i(G)$, and if G is nilpotent, of class c, then $\gamma_i(G)$ is generated by $\bigcup_{i \leq j \leq c} S_j$ (Corollary 2.17). An easy induction shows that the length of the elements of S_i is at most $f(i) := 3 \cdot 2^{i-1} - 2$.

The reduction from soluble to polycyclic is effected by means of the following:

Proposition 5.8 *Let G be a finitely generated abelian-by-(nilpotent of class c) group. Assume that each 2-generated abelian-by-cyclic subgroup of G of depth $f(c+1)$ is polycyclic. Then G is polycyclic.*

Proof Let $A \triangleleft G$ be abelian with a nilpotent of class c factor group. Then $\gamma_{c+1}(G) \leq A$, and we may assume as well that $A = \gamma_{c+1}$. Since finitely generated nilpotent groups are polycyclic (Theorem 2.18), it suffices to show that A is finitely generated. Let S be a finite set of generators of G, order the set $\bigcup_{i \leq c} S_i$ in some way, and consider $x \in S$, $y \in S_{c+1}$. By assumption, $H(x,y) := \langle x, y \rangle$ is polycyclic, and therefore $\langle y \rangle^{H(x,y)}$, which is generated by all the elements $x^{-i} y x^i$, is generated by finitely many of them, say the ones with $|i| \leq k(x, y)$. We let $m = \max\{k(x, y) \mid x \in S, \ y \in S_{c+1}\}$, and claim that A is generated by the elements y^z, with $y \in S_{c+1}$, z a product of elements of $\bigcup_{i \leq c} S_i$, say $z = \Pi z_j^{e_j}$, where the factors z_j occur in the given order, and $|e_j| \leq m$.

To see that, let B be the subgroup generated by that set of conjugates of elements of S_{c+1}. Then $B \leq A$, and B contains the set S_{c+1} of normal generators of A, therefore it suffices to show that $B \triangleleft G$. We will show, by reverse induction on i, that B is invariant under $\gamma_i(G)$, starting with $i = c + 1$. For $i = 1$ that means that B is a normal subgroup.

Assume, then, that we know already that B is $\gamma_{r+1}(G)$-invariant. It suffices to show that B is S_r-invariant. Let $w \in S_r$, let y^z, where $z = \Pi z_j^{e_j}$, be one of the generators of B, and consider y^{zw}. Then $w = z_t$ for some t. Suppose first that $e_t < m$. Then $zw = z'u$, for some $u \in \gamma_{r+1}(G)$, and z' is obtained from z by changing e_t to $e_t + 1$. Then $y^{z'}$ is one of the generators of B, and $y^{zw} = (y^{z'})^u \in B$ by induction.

Let $e_t = m$. Then we write $zw = z_t^{m+1} z' u$, where this time z' is obtained from z by omitting z_t, and $u \in \gamma_{r+1}$. By the choice of m, we have $y^{z_t^{m+1}} = \Pi_{|i| \leq m} (y^{z_t^i})^{f_i}$, for some integers f_i. This exhibits y^{zw} as a product of factors $(y^{f_i})^{z_t^i z' u}$, and now $(y^{f_i})^{z_t^i z'}$ is one of the generators of B, and we can apply induction as before.   QED

This proposition implies an analogue of Proposition 5.6 for a much larger class of groups.

Proposition 5.9 *There exists a constant* $K = K(c) > 1$, *such that if* G *is a finitely generated abelian-by-(nilpotent of class c) group, and* $\Omega(G) < K$, *then* G *is polycyclic.*

Proof Take $K = 2^{\frac{1}{5f(c+1)}}$. Proposition 5.6 and Lemma 5.7 show that the assumptions of Proposition 5.8 are satisfied.   QED

Though we have found an explicit value for K, we did not write it in the formulation of Proposition 5.9, because it is possible that that explicit value is far from the best possible. Now suppose that we want to prove that finitely generated soluble groups are either of uniform exponential growth, or are nilpotent-by-finite. Suppose that G is a finitely generated soluble group of non-uniform exponential growth. Let N be the last non-identity term in the derived series of G. We can employ induction on the derived length to deduce that G/N is nilpotent-by-finite, say that $H \leq G$ is of finite index, and H/N is nilpotent. By Theorem 2.5, H is also of non-uniform exponential growth. Thus Proposition 5.9 shows that H, and with it G, is polycyclic, and it suffices to prove our claim for polycyclic groups.

6
Linear Groups

The term *linear group* refers to a group which is isomorphic to a subgroup of the general linear group $\mathrm{GL}(n, F)$ for some natural number n and some field F. The basic result, settling the growth problem for linear groups, is the following

Theorem 6.1 (The Tits Alternative – [Ti 72]) *Let G be a finitely generated linear group. Then either G contains a non-abelian free subgroup, or G contains a soluble subgroup of finite index.*

Corollary 6.2 *The growth of a finitely generated linear group is either exponential or polynomial, and it is polynomial if and only if the group is nilpotent-by-finite.*

Indeed, if the group contains a non-abelian free subgroup its growth is exponential. In the other case, apply Corollary 5.4.

Y. Shalom [Sh 98] gave a proof of Corollary 6.2 independent of the Tits' alternative for the case of linear groups of characteristic zero. This case suffices for the proof of Gromov's theorem described below. It was proved by A. Eskin, S. Mozes, and Hee Oh [EMO 05], that linear groups of characteristic zero of exponential growth are of uniformly exponential growth, and E. Breuillard and T. Gelander [BG 08] extended this to all characteristics. This is done by showing that if G is not nilpotent-by-finite, then we can find a non-abelian free subsemigroup of G, for which the lengths of the free generators are bounded, relative to whatever set of generators of G is taken. The proofs of the uniform exponential growth of soluble groups follow a similar strategy.

For the proof of Gromov's theorem we need also the following:

Theorem 6.3 (Jordan's Theorem) *A finite subgroup of $\mathrm{GL}(n, F)$, where*

F is a field of characteristic 0, has an abelian subgroup of index bounded in terms of n only.

It is remarkable that C. Jordan proved that theorem in 1878, when group theory was still in its infancy. There are several proofs: see, e.g. [Is 76, Theorem 14.12]. In [CR 62, Section V.36] there are proofs of generalizations to infinite groups, and in [Co 07] the best possible bound is found; this is usually $(n + 1)!$, realized by the symmetric group S_{n+1}. The proof of that depends on the classification of finite simple groups. Variations for finite characteristics are given in [LP 98] and [Co 08].

The bridge between these results and the growth of general groups, in particular polynomial growth, was constructed by Gromov, and it relies on:

Theorem 6.4 (Gleason–Montgomery–Zippin: solution of Hilbert's Fifth Problem [Ka 95, MZ 55]) *Let T be a finite dimensional, locally compact, connected and locally connected, homogeneous metric space. Then the group of isometries of T can be given the structure of a Lie group with finitely many components.*

It is possible that the reader is not familiar with all the terms in this theorem, so we will review some of the definitions. We do not recall the full formal definition of Lie groups. Just recall first that a *topological group* is a group on which a topology is defined in such a way that the group operations, i.e. multiplication and inversion, are continuous. Roughly speaking, a Lie group is a topological group in which the underlying topological space is a manifold, i.e. each point has an open neighbourhood homeomorphic to a Euclidean space, and the group operations are not only continuous, but analytic. The properties of such groups that we need below are given by:

Proposition 6.5 *Let G be a Lie group with finitely many components.*

(a) *G has a normal abelian subgroup Z, such that G/Z is isomorphic to a subgroup of $\mathrm{GL}(k, \mathbb{C})$, for some k.*

(b) *For each natural number n there exists an open neighbourhood of the identity in G which does not contain any non-identity element of finite order less than n.*

(In part (a), Z is the centre of the component of the identity, a normal subgroup of finite index).

Next, the space T is *homogeneous* if for any two points $x, y \in T$ there exists an isometry of T moving x to y.

Now we define finite dimension and Hausdorff dimension. A topological space has *dimension* 0, if each point has an open neighbourhood with empty boundary. It has *dimension at most* n, if each point has an open neighbourhood with boundary of dimension at most $n - 1$. The dimension *equals* n, if it is at most n, but it is not at most $n - 1$. It is easy to see that Euclidean n-space has dimension at most n, but not so easy to prove that the dimension is exactly n. But if we consider in that space the set of all points with irrational coordinates, this is a 0-dimensional space. For more details, we refer to [HW 42].

The Hausdorff dimension is a metric, rather than topological, concept. Recall that the measure of a subset of Euclidean space is determined by looking at covers of that subset by small balls, and adding up the volumes of these balls. The volume of a ball, in turn, is proportional to a power of its diameter, and the exponent in this power is the dimension of the space. If we use the wrong exponent, say we take an exponent larger than the dimension, all subsets will have measure 0, while if we employ an exponent smaller than the dimension, most subsets will have infinite measure. The idea of the Hausdorff dimension is to reverse these observations: first define measures for all exponents, and use the vanishing of these measures to detect the exponent that should be considered as the dimension of the space. The exact definition is:

Definition Let T be a metric space. The *Hausdorff dimension* of T is the infimum of all numbers s, such that for each $\epsilon > 0$, T can be covered by a finite or countable collection $\{A_i\}$ of subsets such that A_i has diameter d_i, and $\sum d_i^s < \epsilon$. In other words, the Hausdorff dimension is the infimum of the numbers s such that the "s-dimensional measure" of T is 0.

Unlike the topological dimension, the Hausdorff dimension need not be an integer, and indeed any non-negative real number can occur as the Hausdorff dimension of some space. The connection between the two dimensions is given by:

Theorem 6.6 *The dimension of a metric space T is the infimum of the Hausdorff dimensions of all metric spaces homeomorphic to T.*

Thus the dimension does not exceed the Hausdorff dimension. For the proof, see [HW 42, chapter VII]. We need the following:

Proposition 6.7 *If a metric space T can be covered, for each ϵ, by k*

balls of diameter ϵ, where k may vary with ϵ, but the product $k\epsilon^d$ remains bounded, then the Hausdorff dimension of T is at most d.

Proof Suppose that $k\epsilon^d < C$, and let $e > d$. Then $k\epsilon^e < C\epsilon^{e-d}$, and the right-hand side may be made arbitrarily small, by taking ϵ small enough. This shows that the Hausdorff dimension is not more than e.   QED

We also recall the definition of the topology that makes the isometry group a Lie group. Fix some base point $e \in T$. For any two positive numbers A and ϵ, let $O(A, \epsilon)$ be the set of all isometries σ such that $d(\sigma x, x) < \epsilon$, for all x such that $d(x, e) \leq A$. The sets $O(A, \epsilon)$ are taken to be a basis for the neighbourhoods of the identity in $\mathrm{Isom}(T)$.

7

Asymptotic Cones

Consider the infinite cyclic group $\mathbb{Z}$, represent its elements as the integer points on the real line, and define the distance between two elements as the distance between the corresponding points. It thus becomes a metric space, though not a very interesting one; all distances are integers, and the topology is discrete. Another possibility is to represent $\mathbb{Z}$ by all multiples of the number $1/2$. Though the distances become smaller, they still comprise a discrete set, and the topology remains discrete. The same happens if we represent $\mathbb{Z}$ by the multiples of $1/3$, or of $1/4$, etc. But if we look at all of these representations, we have marked on the real line all rational points, which are a dense subset of the line. It makes sense to consider the full line as a sort of limit of these representations. Similarly, we can represent the free abelian group $\mathbb{Z} \times \mathbb{Z}$ by the so-called lattice points, i.e. all the points in the plane with integer coordinates, or by all points whose coordinates are integral multiples of $1/2$, etc., and consider the full plane as a limit of these representations. A slightly different view-point is to consider the integer points on the line as the Cayley graph of $\mathbb{Z}$. Then the distance is identical with the distance on the graph. Then we take the same graph, but divide the distances by 2, 3, etc. We can apply a similar process to any group. We start with the Cayley graph, with the distance given by the graph structure. Then we divide the distances by 2, by 3, etc. We have a sequence of discrete metric spaces. Gromov's seminal idea was to try and construct a limit space for this sequence, even when the group does not have any a priori representation inside a familiar space. He describes it as a process of getting further and further away from the Cayley graph; the further away we are, the distances on the graph become smaller and smaller. When we are at infinite distance, we can no longer distinguish between nearby points, and the discrete space becomes a nice continuous one.

A group always acts on its Cayley graph, by left multiplication, say, as a group of isometries. It acts also in the same way on the graph with shrunk distances. Gromov was able not only to construct the limiting space, but also to find a limit action. Essentially this makes the group a group of isometries of a nice metric space. When the original group G is of polynomial growth, the limit space is nice enough to allow application of the results of the previous section, and thus reduce to the case of linear groups, proving that G is nilpotent-by-finite.

Roughly speaking, Gromov's construction proceeds by defining a distance between metric spaces, then showing that if we start with the sequence of Cayley graphs of a group of polynomial growth, with distances shrunk as above, one can find a subsequence G_n of these spaces which converges, in the sense that there exists a space T such that the distances between T and G_n tend to 0. Shortly following Gromov's proof, L. van den Dries and A.J. Wilkie published a different construction of the limit space, using ideas of non-standard analysis, in particular ultraproducts. A simplification of their approach was suggested by Gromov, and this is the approach that we will adopt here. We start by recalling the notion of ultrafilter and ultralimits, then define the limit spaces referred to above, which are called *asymptotic cones*, and derive their basic properties.

Definition   Let S be any set. A *filter* on S is a family $\mathcal{F}$ of subsets of S with the following properties:

(i) $\phi \notin \mathcal{F}$ (ϕ is the empty set).
(ii) $A, B \in \mathcal{F} \Rightarrow A \cap B \in \mathcal{F}$.
(iii) $A \in \mathcal{F}, \ A \subseteq B \Rightarrow B \in \mathcal{F}$.

A maximal filter is called an *ultrafilter*.

An example of an ultrafilter is the family of all subsets containing a fixed element $s \in S$; this is termed a *principal ultrafilter*. If S is finite, all ultrafilters are of this type, but if S is infinite, the family of all complements of finite subsets is a filter, and Zorn's lemma shows that there are ultrafilters containing this filter, and these are not principal. Indeed all non-principal ultrafilters contain this filter of co-finite subsets, and this filter is the intersection of all these ultrafilters. As other examples, consider in any infinite S the complements of all subsets of cardinality smaller than $|S|$, or, on the real line, all complements of sets of finite upper measure, etc. Generally speaking, the idea of a filter is that it contains all large subsets, where the exact meaning of "large" is determined by the context. An ultrafilter divides all subsets into "large" and "small", with no intermediate size recognized. While it is known that there are many ultrafilters, if S is infinite their number is $2^{2^{|S|}}$, no explicit construction for any non-principal ultrafilter is known.

The following simple properties of ultrafilters are very useful. The proofs are left to the reader.

(a) A filter $\mathcal{F}$ is an ultrafilter iff it satisfies: *whenever $A \subseteq S$, then either A or the complement $S - A$ is in $\mathcal{F}$.*

(b) If $\mathcal{F}$ is an ultrafilter, and $S = A_1 \cup \cdots \cup A_n$, then $A_i \in \mathcal{F}$ for some i.

(c) If $\mathcal{F}$ is an ultrafilter, $T \in \mathcal{F}$, and $T = A_1 \cup \cdots \cup A_n$, then $A_i \in \mathcal{F}$ for some i.

(d) If G is a family of non-empty subsets of S, such that the intersection of any finite number of members of G is non-empty, then G is contained in some filter.

(Hint for **(d)**: Consider the family of all subsets containing an intersection of finitely many members of G.)

In the following we are going to use the power A^S of the set A. This is most conveniently considered as the set of functions from S to A.

Definition Let $\mathcal{F}$ be a filter on S, and let A be some set. The *reduced power $A^S/\mathcal{F}$ of A mod $\mathcal{F}$* is the set of equivalence classes in the cartesian power A^S, where two elements $f, g \in A^S$ are *equivalent*, if $\{s \in S \mid f(s) = g(s)\} \in \mathcal{F}$. In words: two functions from S to A are equivalent if they agree on a set belonging to $\mathcal{F}$. Two such functions are said to be equal *almost everywhere* (a.e.).

If $\mathcal{F}$ is an ultrafilter, the reduced power mod $\mathcal{F}$ is called an *ultrapower*.

In other words, the reduced power is the set of functions from S to A, where we identify two functions if they agree on a large subset of S. If A has some structure, say it is a topological space, or a group, a ring, etc., the cartesian power usually has a similar structure, and often the reduced power can also be given the same type of structure. Ultrapowers inherit from A the so-called *elementary properties*. For example, the property of being an invertible element in a ring is an elementary property, therefore if A is a field, then so are its ultrapowers, while the cartesian powers, and reduced powers that are not ultrapowers, have only the structure of rings.

We now fix a non-principal ultrafilter $\mathcal{F}$ on $\mathbb{N}$, the set of natural numbers. Let T be any topological space, and let $\{x_n\}$ be a sequence in T.

For each $x \in T$, and each neighbourhood U of x, write $O(x, U) = \{n \in \mathbb{N} \mid x_n \in U\}$.

Definition   We say x is the $\mathcal{F}$-*limit* of $\{x_n\}$, if for each U the subset $O(x, U)$ belongs to $\mathcal{F}$. We write $x = \mathcal{F} \lim x_n$.

Thus x is the $\mathcal{F}$-limit of some sequence, if each neighbourhood of x contains almost all members of that sequence.

Exercise 7.1 x is the limit of $\{x_n\}$, in the ordinary sense, iff x is the $\mathcal{F}$-limit of $\{x_n\}$, for all non-principal ultrafilters $\mathcal{F}$.

Proposition 7.1

(a) *If T is a Hausdorff space, the $\mathcal{F}$-limit is unique.*
(b) *If T is compact, then each sequence $\mathcal{F}$-converges.*

Proof   (a) Let x and y be two distinct points, and let U and V be disjoint neighbourhoods of them. Then the sets $O(x, U)$ and $O(y, V)$ are disjoint, and at most one of them can belong to $\mathcal{F}$.

(b) Suppose that $\{x_n\}$ does not converge. Then for each point $y \in T$ there is a neighbourhood U_y such that $O(y, U_y) \notin \mathcal{F}$. Then $O^*(y, U_y) := \{n \in \mathbb{N} \mid x_n \notin U_y\} \in \mathcal{F}$. Given any finite number of points $y_1, \dots, y_k$, the intersection $O^* = \bigcap O^*(y_i, U_{y_i}) \in \mathcal{F}$, therefore that intersection is not empty, and if $i \in O^*$, then x_i is outside all of the neighbourhoods U_{y_i}, therefore T cannot be covered by any finite number of these neighbourhoods, contradicting compactness.   QED

Corollary 7.2 *Any bounded sequence of real numbers $\mathcal{F}$-converges, and its $\mathcal{F}$-limit is unique.*

Proposition 7.3 *Let x_n and y_n be two bounded real sequences, and c a real number.*

(a) $\mathcal{F} \lim(x_n + y_n) = \mathcal{F} \lim x_n + \mathcal{F} \lim y_n.$
(b) $\mathcal{F} \lim(c x_n) = c \mathcal{F} \lim x_n.$
(c) *If $x_n \leq y_n$ for all n, then $\mathcal{F} \lim x_n \leq \mathcal{F} \lim y_n.$*

Proof The proof is left to the reader.

Exercise 7.2 A number x is the $\mathcal{F}$ limit of a sequence $\{x_n\}$, for some non-principal ultrafilter $\mathcal{F}$, iff x is the limit, in the ordinary sense, of some subsequence of $\{x_n\}$.

Thus an ultrafilter is a means of choosing, for each sequence, some limit point of it. Moreover, there is some uniformity in this choice. Given two sequences, $\{x_n\}$ and $\{y_n\}$, the $\mathcal{F}$ limits of both can be obtained by using the "same" subsequence of both, i.e. there exists a sequence $\{n_i\}$ of natural numbers, such that $\mathcal{F} \lim x_n = \lim x_{n_i}$ and $\mathcal{F} \lim y_n = \lim y_{n_i}$. This follows from the fact that two sets in $\mathcal{F}$ have non-empty intersection, and actually that intersection is infinite (why?). The uniformity extends to any finite number of sequences.

Exercise 7.3 There is no ultrafilter $\mathcal{F}$ such that $\mathcal{F} \lim x_n = \limsup x_n$, for each real sequence $\{x_n\}$.

We now consider a pair (T, e) of a metric space T and a *base point* $e \in T$. We still have a fixed ultrafilter on $\mathbb{N}$.

Definition A sequence $\{x_n\}$, $x_n \in T$ is *moderate* if it satisfies $d(x_n, e) \leq An$, for some constant A (depending on the given sequence).

Thus a moderate sequence is one which can go to infinity, but in a controlled manner.

Definition Given two moderate sequences, say $\alpha = \{x_n\}$ and $\beta = \{y_n\}$, the *distance* between them is

$$d(\alpha, \beta) = \mathcal{F} \lim(d(x_n, y_n)/n).$$

It is easy to see that this distance satisfies the triangle inequality. However, it is possible for two distinct sequences to be at a zero distance. Therefore we call two sequences *equivalent*, if their distance is zero. This is an equivalence relation, and we define the distance between two equivalence classes as the distance between any two of their representatives (it is obvious that the distance does not depend on the choice of the representatives). The metric space K obtained in this way is termed the *asymptotic cone* defined by (T, e) and $\mathcal{F}$. Note that two sequences that give rise to the same element in the ultrapower of T (mod $\mathcal{F}$), i.e. that are equal on a set in $\mathcal{F}$, are equivalent. That means that we can change the elements of any sequence on a set outside $\mathcal{F}$ without changing the element of K that that sequence represents. But two sequences may be equivalent without being identical on a set of $\mathcal{F}$. We may say that two sequences are equal in K if they are "infinitesimally near" each other.

Remark More generally, instead of requiring $d(x_n, e) \leq An$ we can ask that $d(x_n, e) \leq f(n)$, for some function f, and in the definition of the distance divide by $f(n)$, but the above definition suffices for us.

Let G be a finitely generated group, fix some finite set X of generators of G, let $l(-)$ be the length function with respect to X, and consider G as a metric space, with distance $d(x, y) = l(x^{-1}y)$. Take the identity e of G as the base point, and let K be the asymptotic cone defined by (G, e) and $\mathcal{F}$. We let e denote also the element of K defined by the constant sequence $\{e\}$. Note that the set M of moderate sequences is a group, the set M_0 of sequences equivalent to e is a subgroup, and there is a 1–1 correspondence between K and cosets of M_0 in M. However, M_0 is not a normal subgroup, and therefore K does not carry a natural structure of a group. Nevertheless, it has many nice properties as a metric space.

Exercise 7.4 Let two asymptotic cones be defined by means of the same group and the same ultrafilter, but two different sets of generators. Show that the two cones are quasi-isometric.

Theorem 7.4 *The asymptotic cone K of a finitely generated group G is homogeneous, arcwise connected, locally connected, and complete.*

Proof As in any group, left multiplication (by some fixed element) in M is a permutation, and in this case it is also an isometry, and induces an isometry on K. In particular, if two elements $a, b \in K$ are represented by the sequences $\{x_n\}$ and $\{y_n\}$, then the map sending each element of K, represented by $\{z_n\}$, say, to the element represented by $\{y_n x_n^{-1} z_n\}$, is an isometry of K moving a to b. Thus K is homogeneous.

We will prove that K is arcwise connected by showing that each element a can be connected by a continuous path to e. Recall that this means that there exists a continuous function $f : [0, 1] \to K$, such that $f(0) = e$ and $f(1) = a$. Let a be represented by a sequence $\{x_n\}$ as above, and for each n choose a representation of x_n as a product of length $l(x_n)$ of the generators. Let $0 \le \alpha \le 1$, and for each word w of length k in the generators X, write $w(\alpha)$ for the word consisting of the first $\lceil k\alpha \rceil$ letters in w. Define $f(\alpha)$ as the element represented by the sequence $\{x_n(\alpha)\}$. It is clear that $f(0) = e$ and $f(1) = a$. Moreover, if $0 \le \alpha \le \beta \le 1$, it is also clear that $(\beta - \alpha)l(x_n) - 1 \le d(w(\alpha), w(\beta)) \le (\beta - \alpha)l(x_n) + 1$. Since $l(x_n) \le An$, this implies that $d(f(\alpha), f(\beta)) \le A(\beta - \alpha)$, and therefore f is continuous. This shows that K is arcwise connected, hence connected. Moreover, $d(e, f(\beta)) \le \beta d(e, a)$, and thus the path from e to a is contained in the ball of radius $l(a)$ around e. This shows that each ball around e is arcwise connected. By homogeneity, this holds for all balls, hence K is locally connected.

For the proof that K is complete we need the following remark.

Suppose that a and b are two points of K and that $d(a, b) < A$, for some number A. Then we can find representative sequences $\{x_n\}$ and $\{y_n\}$ for a and b, such that the inequality $d(x_n, y_n) < An$ holds on some set of integers S in the ultrafilter $\mathcal{F}$. For $n \notin S$ we can replace y_n by x_n. That does not change the point b, but ensures that the inequality above holds for all n. Now let $\{a_i\}$ be a Cauchy sequence in K, and let $\{x_{in}\}$ be a representative sequence for a_i. We may assume that $d(a_1, a_i) < 1$ for all i. Changing as above the sequences x_{in}, for $i > 1$, allows us to assume that for each n and each i we have $d(x_{1n}, x_{in}) < n$, and therefore $d(x_{in}, x_{jn}) < 2n$, for all pairs i, j. Let k be the first index such that $k > 1$ and $d(a_i, a_j) < 1/2$ for all $i, j \geq k$. Then change x_{in}, for $i > k$, to obtain that $d(x_{kn}, x_{in}) < n/2$ holds for all $i > k$. This does not change the previous inequalities, because at worst we replaced x_{in} by x_{kn}, which still satisfies $d(x_{1n}, x_{kn}) < n$. Continuing in this way, we see that it is possible to choose the sequences x_{in} in such a way that for each n and each $\epsilon > 0$, the inequality $d(x_{in}, x_{jn}) < \epsilon n$ holds, if i and j are large enough. Fix some n, and then choose ϵ so that $\epsilon n < 1$. Then x_{in} and x_{jn} are at distance less than 1, i.e. there are equal. Therefore the sequence $\{x_{in}\}$ is eventually constant. Write x_n for the eventual value of $\{x_{in}\}$. If m is an index such that $d(x_{in}, x_{jn}) < \epsilon n$ whenever $i, j \geq m$, then in particular $d(x_n, x_{jn}) < \epsilon n$ for all $j \geq m$ (and for all n). Let a be the point of K represented by the sequence $\{x_n\}$. Then the last inequality implies $d(a, a_j) \leq \epsilon$. Thus $\{a_i\}$ converges to a, and the proof of Theorem 7.4 is complete.   QED

The next task is to find actions of the group G on its asymptotic cones. Let $x \in G$. For each sequence $\{x_n\}$ such that $l(x_n) \leq An$ we have $l(xx_n) \leq An + l(x)$, therefore the sequence $\{xx_n\}$ is also moderate, and the map $\{x_n\} \to \{xx_n\}$ is an isometry of M. All isometries of M preserve equivalence, and therefore induce isometries of K. This defines an action of G on K, i.e. a homomorphism $\Phi : G \to \mathrm{Isom}(K)$. For $x \in G$ and $\alpha \in K$ we write $x\alpha$ for $\Phi(x)(\alpha)$, and let $N = \mathrm{Ker}(\Phi)$.

Theorem 7.5 *Let G be a finitely generated infinite group, let K be an asymptotic cone of G, and let $I := \mathrm{Isom}(K)$ be the isometry group of K. Then there exists a homomorphism $\Phi : G \to I$ with kernel N such that one of the following holds:*

(i) *G/N is infinite.*

(ii) *N is abelian-by-finite*

(iii) *For each neighbourhood O of the identity in I there exists a homo-*

morphism $\varphi_O : N \to I$, such that $\mathrm{Im}\,(\varphi_O) \cap O$ *contains non-identity elements.*

The proof requires several steps. First, the homomorphism Φ and its kernel N have been defined already. In order to analyze this action, let us define for each $x \in G$ the *displacement* of x by: $D(x,r) = \max d(a, xa) = \max l(a^{-1}xa)$ (for $a \in G$, $l(a) \leq r$). Here r can be any natural number. If $x \in H \leq G$, and we restrict a above to lie in H, we write $D_H(x,r)$ for the corresponding maximum.

Proposition 7.6 *If $x \in N$, then $\mathcal{F}\lim_{r \to \infty} D(x,r)/r = 0$.*

Proof For each r, choose an a_r such that $l(a_r) \leq r$ and $l(a_r^{-1}xa_r) = D(x,r)$. The sequence $\alpha := \{a_r\}$ is moderate. If $x \in N$, then $x\alpha = \alpha$: therefore $\mathcal{F}\lim D(x,r)/r = d(x\alpha, \alpha) = 0$.   QED

Proposition 7.7 *The function $D(x,r)$ is bounded if and only if x has only finitely many conjugates in G, and in that case $x \in N$.*

This is clear from the definition. Elements having only finitely many conjugates are termed *FC-elements*, and it is clear that they constitute a normal subgroup of G (sometimes called the *FC-centre* of G).

Suppose that $|G : H| < \infty$ and $x \in H$. Then x has finitely many conjugates in G iff it has finitely many conjugates in H, i.e. $D(x,r)$ is bounded iff $D_H(x,r)$ is bounded.

Proposition 7.8 *For $x, y \in G$ and integers r, s we have $D(x, r+s) \leq D(x,r) + 2s$ and $D(y^{-1}xy, r) \leq D(x,r) + 2l(y)$.*

Proof Let $l(a) \leq r + s$. Then we can write $a = bc$ with $l(b) \leq r$ and $l(c) \leq s$. Then $d(xa, a) = d(xbc, bc) \leq d(xbc, xb) + d(xb, b) + d(b, bc) = d(xb, b) + 2d(b, bc) \leq d(xb, b) + 2s \leq D(x,r) + 2s$, hence $D(x, r+s) \leq D(x,r) + 2s$. Next, if $l(a) \leq r$, then $d(y^{-1}xya, a) = d(xya, ya) \leq D(x, r + l(y))$, and we apply the previous inequality.   QED

We need a well-known result of B.H. Neumann:

Proposition 7.9 *A group G cannot be the union of finitely many cosets of subgroups of infinite index.*

Proof Suppose that G is such a union, and that the cosets belong to k distinct subgroups. We use induction on k, the case $k = 1$ being obvious. For $k > 1$, let H be one of the subgroups involved. Since $|G : H| = \infty$, some coset Hx does not occur in the union, and since it is disjoint from the cosets of H that do appear, it is contained in the union of the cosets of the other subgroups. Any coset Hy can be written as $Hx \cdot x^{-1}y$, and this shows that Hy is also contained in a finite union of cosets of the other subgroups. This implies that all cosets of H occurring in the union are contained in finite unions of cosets of the other subgroups, and thus G is the union of finitely many cosets of the other $k - 1$ subgroups, contradicting the inductive hypothesis. QED

Proof of Theorem 7.5. We have already defined Φ and N. Obviously we may assume that $|G : N|$ is finite. Then N is finitely generated, say $N = \langle y_1, \ldots, y_d \rangle$. If $D(y_j, r)$ is bounded, for each j, then $|N : \bigcap_1^d C_N(y_j)|$ is finite. But the indicated intersection is the centre of N, hence (ii) holds. Thus we assume that N contains non-FC elements, and since the FC-elements constitute a proper subgroup, we may assume that none of the generators is an FC-element. Fix some integer r and some ϵ. For each t between 1 and d, the set of elements y of N such that $D(y^{-1}y_ty, r) \leq \epsilon r$ is equal to the set of elements such that $l(a^{-1}y^{-1}y_tya) \leq \epsilon r$ for all $a \in G$ such that $l(a) \leq r$. That means that the conjugate y_t^{ya} of y_t is one of finitely many elements, and therefore the element ya lies in one of finitely many cosets of $C_G(y_t)$, in particular, taking $a = 1$, we see that y itself lies in one of finitely many such cosets. Since each of $y_1, \ldots, y_d$ has infinitely many conjugates in N, i.e. its centralizer has infinite index, the last proposition shows that N is not the union of finitely many cosets of the centralizers of the generators, and so there exists some $z_r \in N$ such that $D(z_r^{-1}y_tz_r, r) > \epsilon r$ for all t. We write z_r as a word in the y_t, and choose the first initial subword x_r of z_r for which $D(x_r^{-1}y_tx_r, r) > \epsilon r$, for some t. We choose for each r one such index $t = t(r)$, and for each $i \leq d$ we write $S(i) := \{r \mid t(r) = i\}$. The finitely many sets $S(i)$ partition $\mathbb{N}$, therefore one of them, say $S(i_0)$, lies in $\mathcal{F}$. Let l be the maximum length, in the generators of G, of the y_t. We may take r to be large enough, and then $D(y_t, r) \leq \epsilon r$, by Proposition 7.6, therefore $x_r \neq 1$, and we can write $x_r = w_ry$, where w_r is the initial subword of z_r preceding x_r, and y is some generator. Then Proposition 7.8 shows that $D(x_r^{-1}y_tx_r, r) \leq D(w_r^{-1}y_tw_r, r) + 2l \leq \epsilon r + 2l$. This holds for each t, but for i_0 we also have $D(x_r^{-1}y_{i_0}x_r, r) > \epsilon r$. It follows that $\mathcal{F}\lim D(x_r^{-1}y_{i_0}x_r, r)/r = \epsilon$. We always have $l(x) \leq D(x, r)$, for any r, and thus the previous inequality shows that $l(x_r^{-1}y_tx_r) \leq \epsilon r + 2l$, for each t, and so, if $y \in N$ has length m in the y_i, we have $l(x_r^{-1}yx_r) \leq m\epsilon r + 2ml$. Therefore left multiplication by the sequence $\{x_r^{-1}yx_r\}$ preserves moderate sequences, and induces an isometry on K. We define $\varphi(y)$ as this isometry. Then φ is a homomorphism $N \to I$.

Applying Proposition 7.6 to this homomorphism shows that $\varphi(y_{i_0}) \neq 1$, because $\mathcal{F}\text{-}\lim D(x_r^{-1} y_{i_0} x_r, r)/r \neq 0$. On the other hand $d(\varphi(y_{i_0})(\alpha), \alpha) \leq \epsilon$ for all $\alpha \in K$, which shows that $\varphi(y_{i_0})$ can be made to lie in any given neighbourhood of the identity in I, by taking a small enough ϵ.   QED

8

Groups of Polynomial Growth

The properties above hold for an asymptotic cone defined by any finitely generated group. We now want to consider some properties which hold only for groups of polynomial growth. Let G be of polynomial growth of degree d. While the sequence $s(n)$ is polynomially bounded, it still can oscillate wildly. We begin by choosing a subsequence with a reasonable behavior, in the sense that we look for those values $s(n)$ whose distance from the earlier terms is not too big.

Proposition 8.1 *Let G be of polynomial growth of degree d. Then there exist infinitely many n such that for all $i < n$ we have (logarithms are to base 2)*

$$\log s(2^n) \leq \log s(2^{n-i}) + i(d+1).$$

Proof By the definition of d we have $\log s_n / \log n < d + 1/2$, for all large enough n. Writing $l(n) = \log(s(2^n))$, we have in particular $l(n)/n < d + 1/2$, therefore $l(n) - n(d+1) < -n/2$, and thus $\lim_{n\to\infty}(l(n) - n(d + 1)) = -\infty$. For each negative integer k let $n(k)$ be the first integer n such that $l(n) - n(d + 1) < k$. Then for $n = n(k)$ and $i < n$ we have $l(n) - n(d + 1) < k \leq l(n - i) - (n - i)(d + 1)$, which is equivalent to the required inequality. Since $n(k)$ takes on infinitely many values, the proposition is proved. QED

Write S for the set of all integers n satisfying the inequalities of the proposition, and $T = \{2^n \mid n + 1 \in S\}$. We choose our ultrafilter $\mathcal{F}$ to contain T. The reason for this particular choice will become clear during the course of the following proof.

Proposition 8.2 *Let G be of polynomial growth, let $\mathcal{F}$ be chosen as just described, and let ϵ be small enough. Then the asymptotic cone K satisfies the following two equivalent properties:*

(1) *If a closed ball of radius 1 in K contains k distinct points, such that the closed balls of radius ϵ around them are disjoint, then $k \leq (1/\epsilon)^{2(d+1)}$.*

(2) *If a closed ball of radius 1 in K contains k distinct points, such that the distances between any two are bigger than 2ϵ, then $k \leq (1/\epsilon)^{2(d+1)}$.*

Proof We first note that the two properties are indeed equivalent. Suppose that (1) holds. Let k points as in (2) be given, then the balls of radii ϵ around them are disjoint, therefore $k \leq (1/\epsilon)^{2(d+1)}$. Conversely, if (2) holds, then, given any k points as in (1), suppose that two of them are at distance $\delta \leq 2\epsilon$. The proof of Theorem 7.4 shows that there exists a continuous path $f(\alpha)$ between these two points, such that $d(f(\alpha), f(\beta)) \leq (\beta - \alpha)\delta$. Then the point $f(1/2)$ lies in the balls of radius ϵ around both points, a contradiction. Therefore (2) implies (1)

Because of homogeneity, it suffices to prove the proposition for the ball B of radius 1 around e. We suppose k points $x_1, \ldots, x_k$ as in (2) are given inside B. Let x_r be represented by the sequence $\{x(r,n)\}$. We may choose the representative sequences so that each point $x(r,n)$ lies in the ball of radius $(1+\epsilon)n$ around e. Choose two distinct indices r and s, and let $N(r,s,\epsilon)$ be the set of ns for which $d(x(r,n), x(s,n)) > 2\epsilon n$. The inequality $d(x_r, x_s) > 2\epsilon$ means that $N(r,s,\epsilon) \in \mathcal{F}$, and then also $N(\epsilon) \in \mathcal{F}$, where $N(\epsilon)$ is the intersection of the sets $N(r,s,\epsilon)$. Let i satisfy $1/2^i < \epsilon \leq 1/2^{i-1}$, and choose $n > i$ such that $2^n \in T \cap N(\epsilon)$. Since $2^n/2^i < 2^n\epsilon$, the balls of radius 2^{n-i} around the points $x(r, 2^n)$ $(r = 1, \ldots, k)$ are disjoint, and each of them holds $s(2^{n-i})$ points, obtaining $ks(2^{n-i})$ points altogether. A point in one of these balls is at distance at most $(1+\epsilon)2^n + 2^{n-i}$ from e, and so these balls lie inside the ball of radius 2^{n+1} around e, and thus we obtain $ks(2^{n-i}) \leq s(2^{n+1})$. But Proposition 8.1 implies $s(2^{n+1}) \leq 2^{(i+1)(d+1)}s(2^{(n+1)-(i+1)})$, and thus $k \leq 2^{(i+1)(d+1)} \leq (1/\epsilon)^{2(d+1)}$. QED

Theorem 8.3 *If G is of polynomial growth, the asymptotic cone K is finite dimensional and locally compact.*

Proof Let us choose inside the ball B the maximal number possible, say k, of points such that the distance between any two of them is more than 2ϵ. Then each point of B lies at distance at most 2ϵ from one of these points, and Proposition 8.2 shows that B is covered by at most $(1/\epsilon)^{2(d+1)}$ balls of radius 2ϵ. This means that the Hausdorff dimension of B is at most $2(d+1)$, and in particular it implies that the dimension of B is finite. Since the dimension is determined by local behavior, the dimension of K is finite. Now let $\{x_n\}$ be any sequence in B. We will show that it has a convergent subsequence. That will show that B is compact, and therefore K is locally compact. To that end, let us cover B, for each i, by k_i balls of radius 2^{-i}. Then an infinite subsequence of $\{x_n\}$ lies in one of the balls of radius 1, an infinite subsequence of it lies in one of the balls of radius $1/2$, etc. Taking a diagonal sequence, we find a Cauchy subsequence of the original sequence. Since K is complete, this subsequence converges. QED

Corollary 8.4 *Let G be an infinite group of polynomial growth. Then there exists a Lie group Γ with finitely many components, and a natural number k, such that G contains a normal subgroup C of finite index, for which one of the following holds:*

(i) *C has an infinite abelian factor group.*
(ii) *C has an infinite factor group in $GL(k, \mathbb{C})$.*
(iii) *There exist homomorphisms $\varphi_n : C \to \Gamma$, for all natural numbers n, such that $|C/Ker(\varphi_n)| \geq n$.*

Proof Let N be as in Theorem 7.5. By Theorems 7.4 and 8.3 the asymptotic cone K satisfies the assumptions of Theorem 6.4. By Proposition 6.5, G/N has a normal abelian subgroup L/N such that G/L is isomorphic to a subgroup of $GL(k, \mathbb{C})$, for some k. If G/L is infinite, we are done, and also if G/L is finite and G/N is infinite. We thus assume that G/N is finite. In case (ii) of Theorem 7.5 we are still done, so we assume that (iii) applies. In that case we apply Proposition 6.5. QED

Now we can deduce:

Corollary 8.5 *An infinite group of polynomial growth contains a subgroup of finite index which has an infinite cyclic homomorphic image.*

Proof Let G be infinite of polynomial growth, and let C be as in Corollary 8.4. In case (ii) of 8.4 we know, by the Tits and Milnor–Wolf theorems, that the infinite linear factor group G/N is nilpotent-by-finite, so G contains a finite index subgroup $H \geq N$ such that H/N is nilpotent and infinite, and the last term in the derived series of H that still has finite index in G is the subgroup that we want. Case (i) is obvious, so we assume that (iii) holds. Let $K_n = Ker(\varphi_n)$. By Theorem 6.5, C contains normal subgroups L_n such that L_n/K_n is abelian and C/L_n is isomorphic to a subgroup of $GL(k, C)$ (where k is independent of n). By Jordan's Theorem 6.3, C contains subgroups $H_n \geq L_n$ such that H_n/L_n

is abelian and $|C : H_n|$ is bounded. Then Proposition 2.3 shows that for infinitely many n the subgroups H_n coincide, equal to a subgroup H, say. Then H/L_n is abelian for infinitely many n, and if the orders of H/L_n tend to infinity, then H/H' is infinite, and H is the subgroup that we need. On the other hand, if the orders H/L_n remain bounded, then for infinitely many n the subgroups L_n coincide, equal to R, say, and then the same argument shows that R is the subgroup that we are looking for.　　　　　　　　　　　　　　　　　　　　　　　　　　　QED

Finally we derive from this Gromov's theorem.

Theorem 8.6 *A group of polynomial growth is nilpotent-by-finite.*

Proof　Let G be a group of polynomial growth, say of degree d. If $d = 0$, then G is finite, so let $d > 0$. By the previous corollary, G contains subgroups $H \triangleright N$ such that $|G : H|$ is finite and H/N is infinite cyclic. By the proof of Theorem 5.1, N is finitely generated, and by Proposition 2.5(d) N has degree of growth $d - 1$ or less, so by induction N contains a nilpotent subgroup K of finite index. By corollary 2.4, we may assume that K is characteristic in N, and hence normal in H. Then H/K contains the finite normal subgroup N/K, with infinite cyclic factor group. Let $H/N = \langle xN \rangle$. Write $C = \langle K, x \rangle$. Then $H/K = C/K \cdot N/K$, hence $|H : C|$ is finite. Since C/K is infinite cyclic, C is soluble, hence nilpotent-by-finite, by Corollary 5.4.　　　　　　　　　　　　　　QED

Corollary 8.7 *A finitely generated group which is quasi-isometric to a virtually nilpotent group is itself virtually nilpotent.*

This follows immediately from Gromov's theorem, because polynomial growth is preserved under quasi-isometry. In contrast, it is known that virtual solubility is not preserved under quasi-isometry [Dy 00].

9

Infinitely Generated Groups

We derive now "finite" versions of Gromov's theorem. These show that it suffices to assume polynomial growth up to some finite length. They also show that if we know the growth function, we can bound the index and the nilpotency class of the nilpotent subgroup. That will enable us to derive versions that apply to infinitely generated groups [Ma 07]. Recall that we saw already in Corollary 4.4 that if G has growth degree d, we can choose the nilpotent subgroup to have class at most $\sqrt{2d}$. The degree does not suffice to bound also the index, but it does so modulo some finite normal subgroup. In particular, if the group is torsion-free, we can bound both the nilpotency class and the index by means of the degree only.

Theorem 9.1 *Let numbers C and d be given. Then there exist numbers k and r, such that if a finitely generated group G satisfies $s_G(n) \leq Cn^d$ for $n = 1, \ldots, k$, then G contains a nilpotent subgroup of class less than $\sqrt{2d}$ and index at most r. The numbers k and r depend only on C and d.*

This was proved already in Gromov's original paper (except for the explicit bound $\sqrt{2d}$), and was applied there to derive the first part of Corollary 9.9 below (again, without the explicit bound $\sqrt{2d}$, and with a worse value for $g(d)$). See [VdDW 84(1), section 7] for a model theoretic proof. Our proof is similar in principle, but we employ only the notion of ultraproducts, avoiding model theoretic notions and results.

Proof Suppose that for some pair C, d the result does not hold. Then we can find groups G_i, $i = 1, 2, \ldots$, satisfying $s_{G_i}(n) \leq Cn^d$, for $n = 1, \ldots, i$, but G_i does not contain a nilpotent subgroup of class less than $\sqrt{2d}$ and index at most i. Substituting $n = 1$ in the inequalities, we see that the growth functions are relative to sets of at most C generators, and without loss of generality we assume that C is an integer, and that each G_i is generated by exactly C elements, say $x_{i,1}, \ldots, x_{i,C}$. Let $\mathcal{F}$ be a non-principal ultrafilter on $\mathbb{N}$, and form the corresponding ultraproduct K of the groups G_i. Let G be the subgroup of K generated by the elements $x_1, \ldots, x_C$, whose representatives are the sequences $\{x_{i,1}\}, \ldots, \{x_{i,C}\}$. Consider $s := s_G(n)$, for some n, and the elements $y_1, \ldots, y_s$ of G of length at most n. Write each y_j as a word w_j of minimal length in the generators. Then for any two distinct indices j, l the set $T(j, l) := \{i \in \mathbb{N} \mid w_j(x_{i,j}) \neq w_l(x_{i,l})\}$ lies in $\mathcal{F}$, and so does their intersection T (the inequalities are in G_i). Therefore T is infinite, and choosing $i \in T$, $i > n$, we see that G_i has at least s distinct elements of length at most n. By assumption, $s \leq Cn^d$, which means that G has polynomial growth, and contains a nilpotent subgroup H of finite index r and class $c < \sqrt{2d}$.

Let F be the free group on C generators $a_1, \ldots, a_C$, and let N be the inverse image of H under the natural homomorphism from F onto G. Then $|F : N| = r$, and in the natural homomorphism of F onto G_i, N maps to a subgroup H_i of index at most r. Let N be generated by elements $\{b_v\}$, $v = 1, \ldots, t$, written as words u_v in the generators. Then in G the elements $u_v(x_j)$ generate H, and each commutator of weight $c + 1$ in these generators is 1. But that means that $\mathcal{F}$ contains the set of indices i for which the same commutators, on the elements $u_v(x_{i,j})$, are 1 (in G_i), and in particular this set of indices is infinite. But the elements $u_v(x_{i,j})$ generate H_i, and therefore for infinitely many is G_i contains a nilpotent subgroup of class at most c and index at most r, contradicting our choice of G_i.   QED

Note that Theorem 9.1 applies also for finite groups, provided that they are large enough to make the statement of the theorem non-trivial. However, we do not know how large that is; for that we need explicit bounds for k and r in terms of C and d. In [VdDW 84(1)] it is shown that there is an algorithm for calculating such bounds, but that algorithm has nothing to do with practical calculation, at least in the present state of computing machinery. We are going to apply Theorem 9.1 in the other direction, for infinitely generated groups. Start with some definitions.

Definition   A group G is *locally of polynomial growth* if each finitely generated subgroup of G is of polynomial growth. G is *locally uniformly of polynomial growth* if there exist numbers C and d, such that each finitely generated subgroup H of G has a set of generators relative to which it satisfies $s_H(n) \leq Cn^d$. It is *locally of polynomial growth of* *degree d*, if each finitely generated subgroup is of polynomial growth of degree *d*.

Definition A group G has *finite rank* if there exists a number r such that each finitely generated subgroup of G can be generated by r elements. The least such r is termed the *rank* of G.

A simple example of a group of finite rank is the additive group of rational numbers. There each finitely generated subgroup is cyclic. It is also clear that that group is locally uniformly of polynomial growth. Direct sums of finitely many copies of that group provide more examples. Together with subgroups of these groups, we have a continuum of examples.

Next we recall the notion of an inverse limit.

Definition A *directed set* is a partially ordered set, in which for any two elements there exists an element bigger than both. An *inverse system* is a family $\{S_\alpha \mid \alpha \in A\}$ of sets, where the index set A is a directed set, and where for each pair $\alpha, \beta \in A$ such that $\alpha \geq \beta$ there exists a map $\phi_{\alpha\beta} : S_\alpha \to S_\beta$, such that if $\alpha \geq \beta \geq \gamma$ then $\phi_{\alpha\gamma} = \phi_{\alpha\beta}\phi_{\beta\gamma}$. The *inverse limit* of the system is the set of all elements $\{s_\alpha\}$ of the cartesian product ΠS_α such that $\phi_{\alpha\beta}(s_\alpha) = s_\beta$ whenever $\alpha \geq \beta$.

If in that definition the sets S_α have some additional structure, e.g. they are topological spaces, or groups, etc., we require the maps $\phi_{\alpha\beta}$ to preserve that structure, i.e. be continuous, or homomorphisms, etc. Usually the inverse limit will have a similar structure. For example, if each S_α is a topological space, then the inverse limit is given a topology as a subspace of the cartesian product with the usual product topology, and if each S_α is a group, then the inverse limit is a subgroup of the cartesian product.

Proposition 9.2 *The inverse limit of a system of compact topological spaces is not empty.*

Proof Let the S_α above be compact spaces. Then their cartesian product S is also compact, by Tychonoff's theorem. For each pair $\gamma, \beta \in A$ such that $\beta \geq \gamma$ let $C_{\beta\gamma}$ be the subset of the elements $\{s_\alpha\}$ of the cartesian product such that $\phi_{\beta\gamma}(s_\beta) = s_\gamma$. This is a closed subset of S. Let $\omega \in A$ be an element bigger than both β and γ. Then any element $\{s_\alpha\}$ of S in which $s_\beta = \phi_{\omega\beta}(s_\omega)$ and $s_\gamma = \phi_{\omega\gamma}(s_\omega)$ belongs to $C_{\beta\gamma}$. Thus these sets are not empty, and, by compactness, neither is their intersection. But that intersection is exactly the inverse limit.   QED

Corollary 9.3 *The inverse limit of a system of non-empty finite sets is not empty.*

Exercise 9.1. Find an inverse system of non-empty sets with an empty limit.

Theorem 9.4 *A group G is locally uniformly of polynomial growth if and only if G is nilpotent-by-finite and of finite rank.*

Proof Let G be locally uniformly of polynomial growth, and let C and d be as in the definition of that term. We remarked already that each finitely generated subgroup of G is generated by C elements. By Theorem 9.1, there exists a number k such that each finitely generated subgroup H contains a normal nilpotent subgroup N of class less than $\sqrt{2d}$ and index at most k. Let $A(H)$ be the set of such subgroups of H, and for each pair $H \leq K$ of finitely generated subgroups, consider the map $N \to N \cap H$ from $A(K)$ into $A(H)$. This defines an inverse system of finite sets and maps, and there exists a point in the inverse limit of that system. Let $\{N(H)\}$ be a point in the inverse limit, and write $N = \bigcup N(H)$. If $x, y \in N$, then $x \in N(H)$ and $y \in N(K)$ for some finitely generated subgroups H and K. Write $L = \langle H, K \rangle$. Then $x, y \in N(L)$, therefore $xy \in N(L) \leq N$, and thus N is a subgroup, and a similar argument shows that it is normal. Let $c = \lfloor \sqrt{2d} \rfloor$, and let $x_1, \ldots, x_{c+1} \in N$, say $x_i \in N(H_i)$. Then $x_i \in N(L)$, where this time L is generated by all the H_i, therefore $[x_1, \ldots, x_{c+1}] = 1$, and N is nilpotent of class c. Finally, if $z_1, \ldots, z_{k+1} \in G$, and $L = \langle z_1, \ldots, z_{k+1} \rangle$, then two of the z_i lie in the same coset of $N(L)$, and therefore in the same coset of N. Thus $|G : N| \leq k$.

For the converse, suppose that G is nilpotent-by-finite and of finite rank, and that each finitely generated subgroup of G can be generated by r elements. Let H be a nilpotent subgroup, say of class c, of finite index k. Let F_r be the free group of rank r, and let N be the intersection of the kernels of all homomorphisms of F_r into G. For each such homomorphism, with kernel K, say, let L be the inverse image of H. Then $|F_r : L| \leq k$ and $\gamma_{c+1}(L) \leq K$. Let M be the intersection of all subgroups of index at most k in F_r. This is a subgroup of finite index, and $\gamma_{c+1}(M) \leq K$, for each K, therefore $\gamma_{c+1}(M) \leq N$. Thus F_r/N is nilpotent-by-finite. Let S be any finitely generated subgroup of G. Then S can be generated by r elements, hence it is a homomorphic image of F_r. In this homomorphism N is in the kernel. Thus S is an image of F/N, and the growth function of F/N (which is polynomial) is an upper bound for the growth function of S. Thus all finitely generated subgroups of G have a common polynomial upper bound for their growth functions. QED

The remaining problem is to study nilpotent groups of finite rank. To give some examples of such groups, recall that if R is a commutative ring, then $U(n, R)$: the group of upper unitriangular n-by-n matrices over R, is nilpotent (see the beginning of Section 2.4; we apply the notation there).

Proposition 9.5 *The group $U := U(n, \mathbb{Q})$ is nilpotent of finite rank.*

Proof We have already noted that U is nilpotent. Let H be a finitely generated subgroup of U, and let $H_i = H \cap U_i$. Since H is nilpotent, its subgroups are also finitely generated In particular, H_i/H_{i+1} is a finitely generated subgroup of the direct sum of $n - i$ copies of the additive group of $\mathbb{Q}$, and therefore it can be generated by $n - i$ elements. It follows that H can be generated by $n(n + 1)/2$ elements. QED

This proposition has a converse: *a torsion-free nilpotent group of finite rank is isomorphic to a subgroup of $U(n, \mathbb{Q})$ for some n* (see the Corollary on p. 25 of [We 73] for a proof). Moreover, a finite extension of a $\mathbb{Q}$-linear group is $\mathbb{Q}$-linear ([We 73], Lemma 2.3). Thus groups of locally uniformly polynomial growth are linear over $\mathbb{Q}$.

We now turn to results in which we restrict only the growth degree. First, two auxiliary results, which will be proved later.

Theorem 9.6 *The orders of the finite subgroups of $\mathrm{GL}(n, \mathbb{Z})$ are bounded by some function $g(n)$ of n alone.*

The next result is a generalization of Theorem 3.10.

Theorem 9.7 *If $|G : Z_n(G)|$ is finite, then $\gamma_{n+1}(G)$ is finite.*

Theorem 9.8 *Let G be a finitely generated group of polynomial growth of degree d. Then G contains normal subgroups N and L, such that $L \leq N$, L is finite, N/L is nilpotent, $\mathrm{cl}(N/L) < \sqrt{2d}$, and $|G : N| \leq g(d)$, where $g(n)$ is the function of Theorem 9.6.*

Proof We know already that G contains a normal nilpotent subgroup K of finite index and of class less than $\sqrt{2d}$. Consider the central series $\delta_i(K)$, and for each $i \leq c := \mathrm{cl}(K)$ write $C_i = C_G(\delta_i(K)/\delta_{i+1}(K))$, and let $N = \bigcap C_i$. Then $K \leq N$. G/N is the group of automorphisms that G induces on $\bigoplus \delta_i(K)/\delta_{i+1}(K)$, and this direct sum is a free abelian group, of rank at most d, by Theorem 4.2. Thus $|G : N| \leq g(d)$, by Theorem 9.6. By definition, we have $K/\delta(K) \leq Z_c(N/\delta(K))$, and therefore $\gamma_{c+1}(N/\delta(K))$ is finite, by Theorem 9.7. But $\delta(K)$ is finite, hence $L := \gamma_{c+1}(N)$ is finite as well, and we are done. QED

If G is as in this theorem, and H is finite, then $G \times H$ also satisfies the assumptions of the theorem. That means that we cannot put any restrictions on L in the theorem. On the other hand we have:

Corollary 9.9 *Let G be a finitely generated torsion-free group of polynomial growth of degree d. Then G contains a nilpotent normal subgroup of class less than $\sqrt{2d}$ and index at most $g(d)$. Moreover, G can be generated by k elements, where k is a number depending only on d.*

Proof The first part is an obvious consequence of Theorem 9.8, and the second one follows, since we have seen in Corollary 4.4 that N can be generated by d elements, and G/N can be generated by $\log_2 g(d)$ elements. QED

For $d = 1$ the last two results were proved before, as Theorem 3.7 and Corollary 3.8.

Before formulating the next result, recall from Section 2.3 that a *Noetherian group* is one satisfying the maximum condition. G is *locally Noetherian* if each finitely generated subgroup is Noetherian.

Theorem 9.10 *The following conditions on a group G are equivalent:*

(1) *G is locally of polynomial growth of degree d, for some d.*
(2) *G is locally Noetherian, and contains normal subgroups N and L such that $L \leq N$, L is locally finite, N/L is a torsion-free nilpotent of finite rank, and $|G : N|$ is finite.*

> *Moreover, if* **(1)** *holds, then N and L can be found such that N/L has class less than $\sqrt{2d}$, rank at most d, and $|G : N| \leq g(d)$. If in* **(2)** *N/L has class c and rank r, then* **(1)** *holds with $d \leq rc$.*

Proof **(1)** implies **(2)**. By Theorem 9.8, each finitely generated subgroup H contains a subgroup $N(H)$ of index at most $g(d)$ such that $\gamma_{c+1}(N(H))$ is finite, where $c = \lfloor \sqrt{2d} \rfloor$. Let $A(H)$ be the set of such subgroups, and for each pair $H \leq K$ of finitely generated subgroups, consider the map $N \to N \cap H$ from $A(K)$ into $A(H)$. This defines an inverse system of finite sets and maps, and there exists a point in the inverse limit of that system. If $\{N(H)\}$ is that point, then, as in the proof of Theorem 9.4, $N := \bigcup N(H)$ is a normal subgroup of index at most $g(d)$. A finitely generated subgroup of $\gamma_{c+1}(N)$ is contained in $\gamma_{c+1}(N(H))$, for some finitely generated subgroup H, and therefore $\gamma_{c+1}(N)$ is locally finite. $\delta(N/\gamma_{c+1}(N))$ is also locally finite, and we take L to be the subgroup of G satisfying $L/\gamma_{c+1}(L) = \delta(N/\gamma_{c+1}(N))$. Each finitely generated subgroup H of N has Hirsch length at most d, by Theorem 4.2, and looking in the series $\delta_i(H/H \cap L)$, with free abelian factors, shows that $H/H \cap L$ can be generated by d elements. Finally, G is locally Noetherian, because each finitely generated subgroup is nilpotent-by-finite.

(2) implies **(1)**. Let H be a finitely generated subgroup of G. Since H is Noetherian, $H \cap L$ is finitely generated, and since L is locally finite, $H \cap L$ is finite. Then the series $H \cap L \triangleleft H \cap N \triangleleft H$, with the first and last links finite and the middle one nilpotent, shows that H has polynomial growth. Moreover, that middle link has a central series of length c with factors that can be generated by r elements, therefore 4.2 shows that the degree of H is at most cr. QED

Corollary 9.11 *Let G be a torsion-free group, which is locally of polynomial growth of degree at most d. Then G contains a nilpotent subgroup of finite index and rank. Moreover, the nilpotent subgroup can be chosen to have class less than $\sqrt{2d}$, rank at most d, and index at most $g(d)$.*

We now prove the auxiliary results.

Proof of 9.7. By induction on n, the case $n = 1$ being Schur's theorem 3.10. Let $n > 1$. By induction, applied to $G/Z(G)$, $\gamma_n(G)Z(G)/Z(G)$ is finite. By 3.10, $\gamma_n(G)'$ is finite. Dividing by the last subgroup, we may assume that $\gamma_n(G)$ is abelian. Since $[\gamma_n(G), Z_n(G)] = 1$, $\gamma_{n+1}(G)$ is generated by the finitely many commutators of representatives of G mod $Z_n(G)$ with representatives of $\gamma_n(G)$ mod $Z(G)$. Thus $\gamma_{n+1}(G)$ is a finitely generated abelian group. Let $[y, x]$ be a typical generator, with $y \in \gamma_n(G)$. The mappings $z \to z^{-1}$ and $z \to z^x$ are endomorphisms of the abelian group $\gamma_n(G)$, hence so is their sum $z \to [z, x]$. For some k we have $y^k \in Z(G)$, therefore $[y, x]^k = [y^k, x] = 1$. Thus the generators of $\gamma_{n+1}(G)$ have finite orders, and so $\gamma_{n+1}(G)$ is finite. QED

Proof of 9.6. Let p be an odd prime, and let σ be the homomorphism from $\mathrm{GL}(n, \mathbb{Z})$ into $\mathrm{GL}(n, p)$, obtained by reducing mod p the entries of an integral matrix. We will show that if G is a finite subgroup of $\mathrm{GL}(n, \mathbb{Z})$, then σ_G is an isomorphism, and hence $|G|$ is bounded by the order of $\mathrm{GL}(n, p)$. If σ is not a 1-1 map, there exists a matrix A of prime order q in G which is mapped onto the identity matrix I. That means that $A = I + pB$, for some integer matrix B. Write $B = dC$, where C is also an integer matrix, and the greatest common divisor of the entries of C is 1. Then we get $I = A^q = (I + pdC)^q = I + qpdC + \frac{1}{2}q(q - 1)p^2 d^2 C^2 + \cdots$ .. It follows that p divides the entries of qC, which means that $p = q$. Then $\frac{1}{2}(q - 1)$ is an integer, and we can divide further to obtain the contradiction that p divides the entries of C.   QED

What are the best possible values for the various constants in Theorem 9.10, as a function of d? In Corollary 4.4 we saw that $d \geq \frac{c(c+1)}{2} + 1$. That inequality cannot be improved. To see that, start with a free abelian group of rank c, generated by $x_1, \ldots, x_c$, and extend it by an element y inducing on it an automorphism fixing x_c and mapping x_i to $x_i x_{i+1}$ otherwise. The resulting group G is nilpotent of class c with $r(1) = 2$ and $r(i) = 1$ for $2 \leq i \leq c$, hence $d = \frac{c(c+1)}{2} + 1$. Moreover, any subgroup of finite index has the same nilpotency class and growth degree.

Exercise 9.2 Prove the last claim about nilpotency classes.

To get the rank of N/L equal to d we have only to consider a free abelian group of rank d. Coming to the index $|G : N|$, let us choose the notation so that $g(n)$ is the best possible bound in Theorem 9.6. Then the semi-direct product of the same free abelian group by a subgroup of order $g(d)$ of $\mathrm{GL}(d, \mathbb{Z})$ shows that $g(d)$ is the best possible bound for the index. Thus the main question is, what is the value of $g(d)$? This was determined only recently, even though Theorem 9.6 was proved by H. Minkowski as early as 1887. Minkowski determined the value of the least common multiple of the orders of finite subgroups of $\mathrm{GL}(n, \mathbb{Z})$ ([Mi 87]; see also [Fe 97]). The value of $g(n)$ is, with a few exceptions, $2^n n!$. This value, which holds for $n \geq 11$ and for $n = 1, 3, 5$, is achieved by the group of monomial integral matrices, i.e. the matrices with just one non-zero value in each row and column, that value being necessarily ± 1. Surprisingly, the proof of that fact, due to W. Feit, requires the classification of the finite simple groups [Fe 95] (see also [Fr 97]). Note that finding the lcm is the same as finding the maximal p-subgroups of $\mathrm{GL}(n, \mathbb{Z})$, which may explain why finding it is an easier task than finding the maximal order.

However, $g(d)$ is not the best bound in Corollary 9.11. Indeed, any maximal finite subgroup A of $\mathrm{GL}(d, \mathbb{Z})$ contains the matrix $-I$. If G is a torsion-free extension of a free abelian group H of rank d by A, then $-I$ acts on H by inversion, and if $x \in G$ induces $-I$ on H, then $x^2 \in H$, hence x both centralizes and inverts x^2, and therefore $x^2 = 1$, and thus G is not torsion-free. If G has degree d and does not contain a free abelian subgroup of rank d, then the nilpotent subgroup N has Hirsch length at most $d - 1$, and then $|G : N| \leq g(d-1) < g(d)$. Thus in either case the index in Corollary 9.11 is strictly less than $g(d)$. The best possible bound in Corollary 9.11 seems not to have been given.

One possible way to find groups of locally polynomial growth may be the following. By a famous theorem of G. Higman, B.H. Neumann, and H Neumann, each countable group is a subgroup of a two-generator group. Let the countable group G be a subgroup of the finitely generated group H. We can count the elements of G according to their length in H. It is then possible that the growth function of G defined in this manner is polynomial, and then Theorem 9.10 applies. Another possible situation is this: suppose that G is a subgroup of an infinitely generated group H, and that a set of generators for H is given. Then we can still define the length of each element relative to these generators. In H there may be infinitely many elements of each length, but it may happen that G contains only finitely many elements of each length. Note that in that case the exponential bound for the growth function no longer applies.

10

Intermediate Growth: Grigorchuk's First Group

R.I. Grigorchuk was the first to show that groups of intermediate growth exist [Gri 80]. We will describe that example. Later we will show that similar constructions yield groups with a continuum of intermediate growth types. Curiously enough, a group very similar to Grigorchuk's was constructed, by a very different method, already by S.V. Aleshin in 1972 [Al 72]. Aleshin's group is commensurable with Grigorchuk's, hence it is also of intermediate growth, but that fact was not noticed at the time (see Section 3 of the next chapter for Aleshin's method).

We consider transformations of the open unit interval (0,1), but to avoid ambiguities in the definitions of these transformations, we remove from that interval all points whose coordinate is rational with denominator a power of 2. Let E denote the identity transformation, and let P denote the transformation interchanging the two halves (0,1/2) and (1/2,1) with each other, i.e. a point x is mapped either to $x + 1/2$ or to $x - 1/2$. Moreover, given any interval I, we let E_I and P_I denote similarly the identity map on I and the interchange of the two halves of I. Usually, the interval I will be clear from the context, so we will write just E and P for E_I and P_I. We regard the unit interval as the disjoint union of the countably many subintervals $(1 - \frac{1}{2^{n-1}}, 1 - \frac{1}{2^n})$, $(n = 1, 2, \ldots)$. We consider a group Γ generated by four transformations a, b, c, and d. Here a is simply the interchange P, applied to the full interval. The other three generators apply to each of the subintervals above either E or P, as follows: b applies P to each of the first two subintervals, then E to the third one, and then repeats the pattern PPE periodically. The generator c applies similarly the periodic pattern PEP, and d applies the pattern EPP.

It is clear that we have $a^2 = b^2 = c^2 = d^2 = 1$ and $bc = cb = d$, $cd = dc = b$, $db = bd = c$. Thus b, c, and d generate a subgroup isomorphic to

$C_2 \times C_2$. That subgroup can be generated by any two of b, c, d, but using all three generators makes the situation more symmetric. When writing any element of Γ as a word in the generators, if any two of b, c, d appear consecutively, we can replace these two by the third one, and thus we can assume that that word has alternating occurrences of a and one of the trio b, c, d. We will always assume that the elements of Γ are written in that form.

Since a swaps the two subintervals $(0,1/2)$ and $(1/2,1)$, while $b, c,$ and d fix both of them, it follows that each element of Γ either fixes or interchanges these subintervals, and the set of all elements fixing them is a normal subgroup H of index 2, consisting of all elements that can be represented as words in which the letter a occurs evenly many times. Recalling the alternating form of these words, the following is obvious.

Lemma 10.1 *H is generated by the elements b, c, d, aba, aca, ada.*

It is clear too that the elements of H can be written as a product in which the generators $\{b, c, d\}$ alternate with the generators $\{aba, aca, ada\}$.

Given any open interval I, it is, like $(0,1)$, a disjoint union of the countably many subintervals consisting of the first (from the left) half of I, then the first half of the remainder, etc. (the end points of these subintervals are removed). Therefore we can define transformations a_I, b_I, c_I, d_I in exactly the same way as for $(0,1)$, and these transformations generate a group isomorphic to Γ. Again we will omit the subscript I when the identity of that interval is clear from the context. We use that convention in the following lemma, which is proved by inspecting the effect of the relevant elements on the two halves of the unit interval.

Lemma 10.2 *The six elements in Lemma 10.1 induce on $(0, 1/2)$ the transformations $a, a, 1, c, d, b$, respectively, and on $(1/2, 1)$ they induce $c, d, b, a, a, 1$, respectively.*

Corollary 10.3 *H induces on each of the two halves a group isomorphic to Γ.*

Corollary 10.4 *Γ is infinite.*

By the previous corollary, the proper subgroup H has epimorphic images isomorphic to Γ. This is impossible in finite groups.

An element $x \in H$ is completely determined by its effect on both halves of the unit interval. Thus, if we denote by x_l and x_r the maps that x induces on the left and right halves, respectively, then the map

$\phi: \ x \to (x_l, x_r)$ is an embedding of H in $\Gamma \times \Gamma$. Lemma 10.2 lists the values of ϕ on the generators of H.

Lemma 10.5 $D := \langle a, d \rangle$ *is the dihedral group of order 8.*

Proof Since $a^2 = d^2 = 1$, D is dihedral, and its order is twice the order of ad. Writing $(ad)^2 = ada \cdot d$, Lemma 10.2 shows that $\phi(ad)^2 = (b, b)$, an element of order 2. Since ϕ is an embedding, we have $(ad)^4 = 1$.   QED

Lemma 10.6 *Let* $B := \langle b \rangle^G$ *be the normal closure of* $\langle b \rangle$ *in* Γ. *Then* $|\Gamma : B| \leq 8$.

Proof Since $\Gamma = \langle a, b, d \rangle$, the factor group Γ/B is generated by the images of a and d, and by the previous lemma these images generate a group of order 8 at most.   QED

Lemma 10.7 $\phi(H)$ *contains the subgroup* $B \times B$ *of* $\Gamma \times \Gamma$.

Proof We have $\phi(ada) = (b, 1)$ and $\phi(d) = (1, b)$. Since H induces Γ on either half of the unit interval, for any $x \in \Gamma$ there are elements of H that are mapped under ϕ to elements of the form (x, y) and (z, x). Conjugating $(b, 1)$ and $(1, b)$ by these elements, we see that both subgroups $(B, 1)$ and $(1, B)$ are contained in $\phi(H)$.   QED

These lemmas yield now a surprising result.

Theorem 10.8 Γ *is commensurable with* $\Gamma \times \Gamma$.

Proof The previous lemma shows that $\phi(H)$ lies between $\Gamma \times \Gamma$ and $B \times B$, and the latter subgroup has a finite index in $\Gamma \times \Gamma$, therefore $\phi(H)$ has finite index in $\Gamma \times \Gamma$. Since ϕ is an isomorphism, and H has a finite index in Γ, the theorem is proved.   QED

Corollary 10.9 Γ *does not have polynomial growth.*

Proof Suppose that Γ has polynomial growth, of degree d, say. Then $\Gamma \times \Gamma$ has degree $2d$ (by Exercise 2.1; alternatively, we can apply Proposition 2.5(d)), hence the theorem shows that $d = 0$, which is not true, since Γ is infinite.   QED

Digressing, let us mention that there exist finitely generated groups G, which are even isomorphic to $G \times G$; see [TJ 74].

Before discussing the growth of Γ further, we derive several other remarkable properties of Grigorchuk's group. The following is a very useful fact.

Lemma 10.10 *Let* $x \in H$, *and write* $\phi(x) = (x_l, x_r)$. *Then* $l(x_l)$ *and* $l(x_r)$ *are at most* $\frac{1}{2}(l(x) + 1)$.

Proof We have already noted that x can be written as an alternating product of elements of the form u and ava, with $u, v \in \{b, c, d\}$. Let the number of these factors be $2k + r$, where $r = 0, 1$. Let us write x as a product of k pairs of such factors, and perhaps one remaining factor, if $r = 1$. Then each pair contributes 4 to $l(x)$, and maps to a product of length 2 (or less) in x_l and x_r. If $r = 1$, the remaining factor has length 1 or 3, and is mapped to either a single generator or to 1. Thus either $l(x) = 4k$, and each of x_l, x_r has length at most $2k$, or $l(x) = 4k + 1$ or $4k + 3$, and x_l and x_r have lengths at most $2k + 1$. This proves our claim. QED

Theorem 10.11 Γ *is a 2-group.*

Proof Let $x \in \Gamma$. We will prove, by induction on $l(x)$, that $x^{2^n} = 1$ for some n. If $l(x) = 1$, then $x^2 = 1$. Assume that $l(x) > 1$. If a word of length $l(x)$ representing x starts with b, say, then bxb is a conjugate of x of the same length or shorter. Similarly if x starts with c or d. Since conjugate elements have the same order, we may assume that x starts with a. If $l(x) = 2$, then x is ab, ac, or ad. We saw already in Lemma 10.5 that ad has order 4. For ac we have $\phi((ac)^2) = (da, ad)$, and thus ac has order 8, and similarly ab has order 16. Let $l(x) \geq 3$. If x also ends with a, then it has the form aya, hence a conjugate of y, which has shorter length. Thus we assume that x starts with a and ends with b, c, or d, and therefore has an even length $2k$. If k is also even, then $x \in H$, and writing $\phi(x) = (x_l, x_r)$, the length of both x_l and x_r is at most $\frac{1}{2}l(x)$, and since the order of x is the maximum of the orders of x_l and x_r, the induction hypothesis applies. There remains the case that $l(x) = 4r + 2$, for some r. In that case $x^2 \in H$, and writing $\phi(x^2) = (y_l, y_r)$, each of y_l, y_r has length at most $l(x)$. Suppose first that x involves the letter d. Since x^2 (as written) has length $2l(x)$ and is periodic with period $l(x)$, the letter d occurs in it at least twice, at positions that differ by $l(x) = 4r + 2$. That means that when we write x^2 as a product of the generators b, c, d, aba, aca, ada, of H, both d and ada occur. Then in y_l the generator d becomes 1, and in y_r the generator ada becomes 1, and so both y_l and y_r have shorter length than x, and by induction x^2 has 2-power order, and the order of x is twice that.

Suppose now that x does not involve d, but does involve c. Then, with the same notation, y_l and y_r either involve d, or that d disappears by cancellations, and then the relevant y has shorter length. Thus either the previous case or the induction applies. Finally, if neither d nor c occurs, then x is a power of ab, and has order dividing 16. Our theorem is proved.   QED

Thus Γ is an infinite, finitely generated torsion group. The existence of such a group was an open problem for many years, and the first examples were constructed by E.S.Golod in 1964 [Go 64]. The Grigorchuk group was first constructed as another example; it appears to be the simplest known. Only later was it noted that it has intermediate growth. A much more difficult problem is to construct an infinite finitely generated group with all elements of bounded order. This was first done by S.I. Adian and P.S. Novikov in 1968 [Ad 75]: for an alternative construction see [Ol 91]. The existence of both types of group is known as the *Burnside problem*. The Grigorchuk group does not answer the more difficult problem: indeed, no groups of intermediate growth whose elements have bounded orders are known.

It is easy to derive from the last proof a bound for the order of an element x in terms of $l(x)$; however, that bound is far from giving the true values.

Recall that if X is a class of groups, then a group G is termed *residually* X if the homomorphisms of G onto X-groups separate points, i.e. given any two distinct elements $x, y \in G$, there exists a homomorphism $\psi : G \to H$, where $H \in X$, such that $\psi(x) \neq \psi(y)$. Since $x \neq y$ is the same as $xy^{-1} \neq 1$, it suffices to separate the identity from non-identity elements. Equivalently, $\bigcap N = 1$, where N runs over the normal subgroups of G with factor groups in X. Another formulation is: G is a subdirect product of X-groups (i.e. a subgroup of a cartesian product of X-groups, which projects onto each factor). We say that G is *residually-p*, if it is residually a finite p-group.

Theorem 10.12 Γ *is residually-2.*

Proof Consider the 2^n subintervals $(\frac{k}{2^n}, \frac{k+1}{2^n})$ of (0,1). It is clear that Γ permutes these subintervals among themselves. Let H_n be the kernel of this action. Then Γ/H_n is finite. Each point in $(0,1)$ is the intersection of all subintervals of the above type that contain it, and therefore $\bigcap H_n = 1$. Thus Γ is residually finite. By the previous result, a finite factor group of Γ is a 2-group.   QED

Theorem 10.13 *The word problem is soluble in* Γ.

Proof Let x be any word in the generators $\{a, b, c, d\}$. Count the number of occurrences of a. If that number is odd, then $x \notin H$, and certainly $x \neq 1$. If $x \in H$, we can evaluate $\phi(x) = (x_l, x_r)$. By Lemma 10.10, if $l(x) > 1$, then both x_l and x_r are shorter words, and by induction we can determine if they are the identity or not, while if $l(x) = 1$, then x is one of b, c, d, and it is not the identity. Since $x = 1$ iff $x_l = x_r = 1$, we are done.   QED

We now return to growth. To show that Γ has intermediate growth, it remains to show that its growth is subexponential. Let us explain the idea of the proof. Since $|\Gamma : H|$ is finite, it suffices to count elements of H. If $x \in \Gamma$ is written as a word in the generators a, b, c, d, it is easy to see that if $x \in H$, then when we write x as a word in the generators of H, generators of the type u and of the type ava alternate, where u, v are among b, c, d. When we write $\phi(x) = (x_l, x_r)$, elements of the type ava are mapped to elements of the type u. This leads to a reduction in length in the passage from x to either x_l or x_r, which is such that even the sum of lengths of x_l and x_r does not exceed $l(x)$ (except possibly for a small additive constant). Moreover, if x involves the generator d, then, since both d and ada induce the identity on one of the half intervals, the sum of the lengths is usually strictly less than $l(x)$. This reduction of length may be insignificant, if there are only a few occurrences of d in x. However, if we are able to apply ϕ twice to x, i.e. if x_l and x_r also lie in H, then on the first application occurrences of c become occurrences of d, which then disappear partially on the second application of ϕ. Finally, if we are able to apply ϕ three times, then in the third application all occurrences of either b, c, or d become trivial on one of the halves. This leads to a considerable reduction of length for all words, which shows that there are many fewer elements of large length than expected.

Let us make the above ideas precise. First, $|\Gamma \times \Gamma : H \times H| = 4$. Therefore $|H : \phi^{-1}(H \times H)| = |\phi(H) : \phi(H) \cap (H \times H)| \leq 4$. Write $K = \phi^{-1}(H \times H)$. Thus $|\Gamma : K| \leq 8$. Let $L = \phi^{-1}(K \times K)$. Similar reasoning shows that $|\Gamma : L| \leq 128$. On L we can apply ϕ three times, and if $x \in L$ then $\phi^3(x) \in \Gamma \times \Gamma \times \cdots \times \Gamma$, with 8 factors in the product. Write $\phi^3(x) = (x_1, \ldots, x_8)$.

Lemma 10.14 *With the above notation, we have $\sum_1^8 l(x_i) \leq \frac{3}{4}l(x) + 8$.*

Proof Repeating Lemma 10.10 three times shows that the sum is bounded by $l(x) + 7$. The number of occurrences of either b, c, or d in x is between $\frac{1}{2}(l(x) - 1)$ and $\frac{1}{2}(l(x) + 1)$. Let us write $l_b(x)$, $l_c(x)$, and $l_d(x)$

for the number of occurrences of each. We saw that on applying ϕ, we get $l(x_l) + l(x_r) \le l(x) + 1$. But each occurrence of d, either as itself or in ada, becomes 1 in either x_l or x_r. Therefore we have to subtract $l_d(x)$ from the sum. Next, each occurrence of c in x becomes d in either x_l or x_r (but not in both). It is possible that on reducing x_l and x_r this occurrence of d disappears, but if it does not, it becomes 1 on applying ϕ again. Thus when applying ϕ^2, we have to subtract also $l_c(x)$, and similarly, on applying ϕ^3, we subtract also $l_b(x)$. But we cannot subtract both $l_c(x)$ and $l_b(x)$, because c and b may cancel out together. For example, suppose that x contains the subword $abadaca$. This becomes, on applying ϕ, $(c \cdot 1 \cdot d, aba) = (b, aba)$. Thus the two occurrences, of b and c, became one occurrence of b. Therefore we can subtract either $l_d(x) + l_c(x)$ or $l_d(x) + l_b(x)$. Since at least one of these is at least $\frac{1}{2}(l_b(x) + l_c(x) + l_d(x))$, and $l_b(x) + l_c(x) + l_d(x) \ge \frac{1}{2}(l(x) - 1)$, our claim follows.   QED

Theorem 10.15 Γ *has intermediate growth.*

Proof In view of Corollary 10.9, it remains to show that Γ is of subexponential growth. Recall that $\omega(\Gamma)$ denotes $\lim s_\Gamma(n)^{1/n}$, and that we have to prove that $\omega(\Gamma) = 1$. Let k be an upper bound for the lengths of a set of representatives of the cosets of L (by the proof of Proposition 2.5, we can take $k = |\Gamma : L|$). Each element $x \in \Gamma$ can be written as $x = yz$, where $y \in L$ and z belongs to our set of representatives. If $l(x) = n$, then $l(y) \le n + k$. Two different elements can give rise to the same element y only if they lie in different cosets. Let us write $s_L^\Gamma(n)$ for the number of elements of L of length (in Γ) at most n. Then $s_\Gamma(n) \le s_L^\Gamma(n+k)|\Gamma : L|$. For $x \in L$, write $\phi^3(x) = (x_1, \ldots, x_8)$ as in Lemma 10.14. Since ϕ^3 is an embedding, x is determined by the elements x_i. Let $l(x) = n$ and $l(x_i) = n_i$. Then $n_i \le n$, therefore the number of possibilities for the octet $(n_1, \ldots, n_8)$ is at most $(n+1)^8$. Given such an octet, the number of possibilities for $(x_1, \ldots, x_8)$ is $\Pi s_\Gamma(n_i)$. Write $\omega = \omega(\Gamma)$. Given that $\epsilon > 0$, we have $\omega^n \le s_\Gamma(n) \le (\omega + \epsilon)^n$, the second inequality assuming that n is large enough. Then there exists a constant A such that $s_\Gamma(n) \le A(\omega + \epsilon)^n$ for all n. Given an octet $(n_1, \ldots, n_8)$ as above, this implies that the number of possibilities for the octet $(x_1, \ldots, x_8)$ is at most

$$\Pi A(\omega + \epsilon)^{n_i} = A^8 (\omega + \epsilon)^{\sum n_i} \le A^8 (\omega + \epsilon)^{\frac{3}{4}n + 8} \le C(\omega + \epsilon)^{\frac{3}{4}n},$$

for some other constant C. Thus

$$\omega^n \le s_\Gamma(n) \le |\Gamma : L| s_L^\Gamma(n+k) \le |\Gamma : L| C(n+k+1)^8 (\omega + \epsilon)^{\frac{3}{4}(n+k)}.$$

Taking nth roots and letting n go to infinity and ϵ go to 0, we have $\omega \le \omega^{\frac{3}{4}}$. This implies that $\omega = 1$. QED

By elaborating the arguments of Theorem 10.15 and Corollary 10.9, we can get a more precise result.

Theorem 10.16 *There exist numbers $0 < \alpha, \beta < 1$, and $A, B > 1$, such that $A^{n^\alpha} \le s_\Gamma(n) \le B^{n^\beta}$.*

Proof By Lemma 10.10, if $x \in H$, then the length of x_l and x_r is at most $\frac{1}{2}(l(x) + 1)$. Repeating this, we have that if $x \in L$, then, for each i, $l(x_i) \le \frac{1}{8}l(x) + 1$. By Lemma 10.14, one of the x_i has length at most $\frac{3}{32}l(x) + 1$. It is convenient to regard $s(n)$ as defined for all positive numbers, not only integers. Thus $s(x) = s(\lfloor x \rfloor)$. If we know which x_i is the short one, there are at most $s_\Gamma(\frac{3}{32}l(x) + 1)$ possibilities for it, and at most $s_\Gamma(\frac{1}{8}l(x)+1)$ for each of the others. Recalling that $s_\Gamma(m+n) \le s_\Gamma(m)s_\Gamma(n)$, we have $s_\Gamma(\frac{1}{8}l(x) + 1) \le s_\Gamma(\frac{1}{32}l(x) + 1)^4$, and thus the number of possibilities for the octet (x_i) is at most $s_\Gamma(\frac{1}{32}l(x) + 1)^{31}$. Since we do not know which x_i is the short one, we have to multiply this estimate by 8. Thus

$$s_\Gamma(n) \le |\Gamma : L| s_L^\Gamma(n+k) \le 8|\Gamma : L| s_\Gamma(\frac{1}{32}(n+k) + 1)^{31}.$$

Iterating this t times, we get

$$s_\Gamma(n) \le 8|\Gamma : L|\left(8|\Gamma : L| s_\Gamma\left(\frac{1}{32}\left(\frac{1}{32}(n+k) + 1 + k\right) + 1\right)^{31}\right)^{31}$$

$$= (8|\Gamma : L|)^{1+31} s_\Gamma\left(\frac{1}{32^2}(n+k) + \frac{k}{32} + \frac{1}{32} + 1\right)^{31^2} \le \cdots$$

$$\le (8|\Gamma : L|)^{1+31+\cdots+31^{t-1}} \times$$

$$s_\Gamma\left(\frac{1}{32^t}(n+k) + (k+1)\left(\frac{1}{32^{t-1}} + \cdots + \frac{1}{32}\right) + 1\right)^{31^t}.$$

We choose $t = \lfloor \log_{32}(n+k)\rfloor$. Then $\frac{n+k}{32^t} < 32$, and the previous inequality shows that for some constants C and B:

$$s_\Gamma(n) \le (8|\Gamma : L|)^{31^t} s_\Gamma(k+34)^{31^t} = C^{31^t}$$

$$\le C^{31^{\log_{32}(n+k)}} = C^{(n+k)^{\log_{32} 31}} = B^{n^{\log_{32} 31}}.$$

This proves the upper bound. The existence of the lower bound follows from Theorem 10.8, and can be considered as a sharpening of Corollary 10.9. The fact that Γ and $\Gamma \times \Gamma$ are commensurable means that their growth functions are equivalent. But if $s(n) = s_\Gamma(n)$ is the growth function of Γ with respect to the set of generators X, then the growth function of $\Gamma \times \Gamma$ with respect to the set $(X \times 1) \cup (1 \times X) \cup (X \times X)$ is $s(n)^2$. The equivalence of these functions means that for some numbers $C, D > 1$ we have $s(n)^2 \le Cs(Dn)$. Then

$$s(n) \ge \frac{1}{C} s\left(\frac{n}{D}\right)^2 \ge \frac{1}{C^3} s\left(\frac{n}{D^2}\right)^4 \ge \cdots \ge \frac{1}{C^{1+2+\cdots+2^{r-1}}} s\left(\frac{n}{D^r}\right)^{2^r}.$$

Let m be such that $s(m) > C$, write $E = \frac{s(m)}{C}$, and take $r = \lfloor \log_D(\frac{n}{m}) \rfloor$. Then $s(n) \ge \frac{1}{C^{2^r}} s(m)^{2^r} = E^{2^r}$. Here

$$2^r > 2^{\log_D(n/m)-1} = \frac{1}{2} \cdot 2^{\frac{\log_2(n/m)}{\log_2 D}} = \frac{1}{2} \cdot (n/m)^{\frac{1}{\log_2 D}} = F \cdot n^{\frac{1}{\log_2 D}},$$

for some F. Thus $s(n) > (E^F)^{n^{\frac{1}{\log_2 D}}}$, establishing the lower bound.

QED

The values of A and B are not very important, because for a fixed α, all the functions A^{n^α}, for all values $A > 1$, are equivalent. On the other hand, if $\alpha \ne \beta$, then A^{n^α} and A^{n^β} are not equivalent. Therefore the interest is in the values of α and β in Theorem 10.16. That theorem is due to Grigorchuk [Gri 84], who gave the value $\alpha = 0.5$ and the value $\beta = \log_{32} 31 = 0.991 \cdots$ derived above.

The best known values, due to L. Bartholdi, are $\alpha = 0.5157 \cdots$ [Ba 01] and $\beta = 0.7674 \cdots$ [Ba 98]. This value of β will be derived below, but for the lower bound we will only derive Grigorchuk's value $\alpha = \frac{1}{2}$. The following is probably a difficult question:

Problem   Does there exist a number γ such that $s_\Gamma(n)$ is equivalent to 2^{n^γ}?

If there is no such γ, then we would like to know the supremum of the numbers α, and the infimum of the numbers β, for which Theorem 10.16 holds.

To proceed further, we need to elaborate on Theorem 10.12. Note that comparing the notation of that theorem to our previous notation, $H_1 = H$. To identify, e.g., Γ/H_2, as a subgroup of the symmetric group on the four subintervals of length $\frac{1}{4}$, we design these subintervals as 1,2,3,4, then the generators of Γ act as follows: $a \to (1,3)(2,4)$, $b \to (1,2)$, $c \to (1,2)$.

This shows that Γ acts as a Sylow 2-subgroup, of order 8, of S_4. We continue this process one further step, identifying Γ/H_3 as a subgroup of order S_8. Numbering the subintervals as 1 to 8, we see that a, b, c, and d act as $(1,5)(2,6)(3,7)(4,8)$, $(1,3)(2,4)(5,6)$, $(1,3)(2,4)$, and $(5,6)$. It follows that Γ/H_3 is a Sylow 2-subgroup T of S_8, of order 128. Let the reader not think that G/H_n is always a Sylow 2-subgroup of S_{2^n} – quite the reverse! It is known that the minimal number of generators of a Sylow 2-subgroup of S_{2^n} is n, therefore Γ, which has three generators, cannot map onto that Sylow subgroup if $n \geq 4$. We needed to know Γ/H_3 for the following proof. Recall that B denotes the normal closure of b.

Lemma 10.17 $|\Gamma : B| = 8$.

Proof We know already that $|\Gamma : B| \leq 8$. To establish the reverse inequality, we will show that in the homomorphism onto $T \cong \Gamma/H_3$, B maps onto a subgroup of order 16 at most. We noted that b maps onto $u := (1,3)(2,4)(5,6)$, and similarly aba maps onto $v := (1,2)(5,7)(6,8)$. To these two elements we add $w := (1,2)(3,4)$ and $t := (5,6)(7,8)$, and consider the subgroup A of S_8 generated by these four elements. We see that, first, t and w commute with each other and with u, v, and hence they generate a central subgroup C of A of order 4. Next, we have $(uv)^2 = wt \in A$, and hence A/C is also of order 4, and $|A| = 16$. Finally, the four generators of A are mapped into A under conjugation by the generators of Γ, and hence $A \triangleleft T$, and therefore B is mapped into A, and thus $|\Gamma : B| \geq |T : A| = 8$. QED

If we recall that the upper bound on the index was established by noting that $\Gamma = B\langle a, d\rangle$, and $D := \langle a, d\rangle$ is a dihedral group of order 8, we see that Γ is a semi-direct product of B by D. Then $\Gamma \times \Gamma$ is a semi-direct product of $B \times B$ by $D \times D$.

Lemma 10.18 $|\Gamma \times \Gamma : \phi(H)| = 8$.

Proof We know already that $\phi(H) \geq B \times B$, and $B \times B$ has index 64. $\phi(H)$ contains also $\phi(c) = (a, d)$ and $\phi(aca) = (d, a)$, and therefore it contains also $(B \times B)\langle(a, d), (d, a)\rangle$. On the other hand, it is easily seen that the six generators of H are mapped by ϕ into $(B \times B)\langle(a, d), (d, a)\rangle$, and thus the latter subgroup equals $\phi(H)$, and since $D^* := \langle(a, d), (d, a)\rangle \cong D$ and $(B \times B) \cap D^* = 1$, we obtain the claimed equality. QED

We now define several mappings. First, there is a natural projection of

$\Gamma = BD$ onto D, mapping a and d to themselves, b to 1 and $c = bd$ to d. Combining this with the automorphism of D that exchanges a and d, we obtain a map $\tau : \Gamma \to D$, such that $\tau(a) = d$, $\tau(b) = 1$, $\tau(c) = \tau(d) = a$. We also consider the map $\psi : x \to (\tau(x), x)$ of Γ into $\Gamma \times \Gamma$. This map is 1–1.

We have $\phi(aca) = (d, a) = \psi(a)$, $\phi(b) = (a, c) = \psi(c)$, $\phi(c) = \psi(d)$, and $\phi(d) = \psi(b)$. Writing $\sigma = \phi^{-1}\psi$, we obtain that σ is an isomorphism between Γ and its subgroup $\langle aca, d, b, c \rangle$, and that $\phi\sigma(x) = (\tau(x), x)$. Note that $\langle aca, b, c, d \rangle \leq H$, and that $\phi(a\sigma(x)a) = (x, \tau(x))$.

Lemma 10.19 *Let* $x \in H$ *and* $\phi(x) = (x_l, x_r)$. *Then*

$$x = a\sigma(x_l)a\sigma(\tau(x_l)^{-1}x_r).$$

Proof Note that the right-hand side has the form $a\sigma(y)a\sigma(z)$, for some y and z, and therefore it lies in H. It suffices to show that applying ϕ to both sides yields the same result. We have, for some w,

$$\phi(a\sigma(x_l)a\sigma(\tau(x_l)^{-1}x_r)) = \phi(a\sigma(x_l)a)(\phi\sigma(\tau(x_l)^{-1}x_r))$$

$$= (x_l, \tau(x_l))(\tau(w), \tau(x_l)^{-1}x_r) = (x_l\tau(w), x_r) = (x_l, x_r)(\tau(w), 1).$$

This implies that $(\tau(w), 1) \in \phi(H)$. But it is clear from the identification of $\phi(H)$ in the proof of the previous lemma, that $\phi(H) \cap (D \times 1) = 1$, and thus $(\tau(w), 1) = 1$, implying the desired equality.   QED

We can now obtain a sort of converse to Lemma 10.10.

Lemma 10.20 *Let* x, x_l, x_r *be as in the previous lemma. Then* $l(x) \leq 4\max(l(x_l), l(x_r)) + 12$.

Proof To begin with, the inequalities $l(\tau(x)) \leq 4$, $l(\sigma(x)) \leq 2l(x) + 1$ hold: the first, because all elements in D have length at most 4. The second inequality is established by induction on $l(x)$, noting that if x ends by b, c, or d, then even $l(\sigma(x)) \leq 2l(x)$. Given these inequalities, our lemma follows from the previous one.   QED

We recall some arguments from Chapter 2. Given a subgroup K of a group G, recall that $s_K^G(n)$ denotes the number of elements of K that have length at most n relative to the generators of G.

Lemma 10.21 *Assume that* $|G : K| < \infty$, *and let* r *be an upper bound for the lengths of a set of representatives of the cosets of* K. *Then*

$$s_G(n - r) \leq |G : K| s_K^G(n) \leq s_G(n + r).$$

Proof The products of the elements of K by the coset representatives have length at most $n + r$. That proves the right-hand inequality. For the other inequality, note that when we write $x \in G$ as a product $x = yz$ with $y \in K$ and z one of the representatives, then $l(y) \le l(x) + r$, and the number of xs is at most the number of ys times the number of representatives. QED

We apply this lemma for the embedding of H in Γ and of $\phi(H)$ in $\Gamma \times \Gamma$. Recall that in the situation of the lemma, we can choose representatives satisfying $r < |G : K|$. Recall also that if $\{a_i\}$ is a set of generators of G, we can take as generators of $G \times G$ the pairs (a_i, a_j), $(a_i, 1)$, and $(1, a_j)$. Then for $x, y \in G$ we have $l_{G \times G}(x, y) = \max(l_G(x), l_G(y))$. It follows that $s_{G \times G}(n) = s_G(n)^2$ and $s_{K \times K}^{G \times G}(n) = s_K^G(n)^2$. In our situation, if $x \in H$ and $\phi(x) = (x_l, x_r)$, then x_l and x_r are uniquely determined by x, and therefore Lemma 10.19 implies that $s_{\phi(H)}^{\Gamma \times \Gamma}(n) \le s_H^\Gamma(4n + 12)$. Putting the various inequalities and equalities together, we have $\frac{1}{2}s_\Gamma(4n + 12 + 1) \ge s_H^\Gamma(4n + 12) \ge s_{\phi(H)}^{\Gamma \times \Gamma}(n) \ge \frac{1}{8}s_{\Gamma \times \Gamma}(n - 7) = \frac{1}{8}s_\Gamma(n - 7)^2$, which can be put in the form $s_\Gamma(n) \ge \frac{1}{4}s_\Gamma(\frac{n}{4} - 11)^2$. Now for any group G and any fixed K there exists a constant C such that $s_G(n + K) \le Cs_G(n)$ (if G is d-generated, take $C = (2d - 1)^K$). Therefore the last inequality implies that there exists a positive constant B such that

$$s_\Gamma(n) \ge Bs_\Gamma(\frac{n}{4})^2.$$

We thus have the same inequality as in the the proof of the second part of Theorem 10.16, with $D = 4$. Since $\log_2 4 = 2$, the argument there implies:

Corollary 10.22 *There exists a number $A > 1$ such that $s_\Gamma(n) \ge A^{n^{\frac{1}{2}}}$.*

To obtain an upper bound improving the one of 10.16, the first stage is introducing *weights*.

Definition Let the group G be generated by the finite set $X = \{x_1, \ldots, x_d\}$. A *weight* w on G is an assignment of a positive number $w(x_i)$, its *weight*, to each generator x_i. We assign to the inverse x_i^{-1} the same weight, and to each word $u(x) = \Pi_j x_{i_j}^{\epsilon_j}$ we assign as weight $\sum_j w(x_{i_j})$. Here $\epsilon_j = \pm 1$. Finally, to each element $x \in G$ we assign as its weight $w(x)$ the minimal weight of all words representing it.

We write $s_{G,w}(n)$ for the number of elements of G whose weight does not exceed n. This is the *weighted growth function* of G.

The following proposition is immediate:

Proposition 10.23 *Any two weighted growth functions on an infinite group are equivalent.*

In particular, all weighted growth functions are equivalent to the ordinary growth function. It turns out that choosing appropriate weights may make it easier to establish properties of the growth function. In this way Bartholdi was able to obtain the values $\alpha = 0.5157$ and $\beta = 0.7674$ in Theorem 10.16.

To avoid ambiguity in the notation $w(x)$, the weights should be chosen so that if one of the generators, say x_i, can be expressed as a word u in the other generators, then the given value $w(x_i)$ does not exceed $w(u)$ as defined above. This is automatically the case if the generating set is irredundant; no proper subset of it generates G. However, our standard set of generators for Γ is redundant. We thus consider weights for Γ which satisfy

$$w(b) \leq w(c) + w(d), \quad w(c) \leq w(d) + w(b), \quad w(d) \leq w(b) + w(c).$$

These inequalities ensure that given $x \in \Gamma$, the weight $w(x)$ is realized when x is expressed in the standard form $au_1 au_2 \cdots au_k$, where, as usual, the elements u_i are among b, c, d, and where the first factor a and the last factor u_k may be absent. Our aim is to find a weight for which an inequality similar to the one of Lemma 10.14 holds already when we apply ϕ once, not three times. That is, we want to find constants A, η, with $0 < \eta < 1$, such that if $x \in H$, then the inequality $w(x_l) + w(x_r) \leq \eta(w(x) + A)$ holds, and we wish to make η as small as possible. We term that inequality the η *inequality*. It turns out to be convenient to take $A = w(a)$. To find the weights, first note that by taking b, c, d for x we obtain

$$w(a) + w(c) \leq \eta(w(a) + w(b)), \quad w(a) + w(d) \leq \eta(w(a) + w(c)),$$

$$w(b) \leq \eta(w(a) + w(d)).$$

Let us assume that these inequalities hold. Take for x the elements aba, aca, ada. Then the left-hand sides of the resulting η inequalities are the same as for b, c, d, while on the right-hand side $w(x)$ increases by $2w(a)$. Thus the η inequality holds in the stronger form $w(x_l) + w(x_r) \leq \eta(w(x) - w(a))$. Any element $x \in H$ can be written as a product of the six standard generators, and the congruence class of $l(x) \bmod 4$ determines if there are more, the same number, or fewer generators of the type u than generators of the type aua. Adding the η inequalities for the generators yields the following:

Claim *If the η inequality holds for b, c, d, then it holds for all $x \in H$. More precisely, if $l(x) = 4r + 3$, then $w(x_l) + w(x_r) \leq \eta(w(x) - w(a))$; if $l(x) = 4r$, then $w(x_l) + w(x_r) \leq \eta w(x)$; and if $l(x) = 4r + 1$, then $w(x_l) + w(x_r) \leq \eta(w(x) + w(a))$.*

Let us write x, y, z, t, for $w(a), w(b), w(c), w(d)$. Then the inequalities above read

$$x + z \leq \eta(x + y), \quad x + t \leq \eta(x + z), \quad y \leq \eta(x + t).$$

Suppose that these inequalities hold for some $0 < \eta < 1$ and positive x_0, y_0, z_0, t_0. Fixing η and $x = x_0$, find y, z, t in the cube $0 \leq y \leq y_0$, $0 \leq z \leq z_0$, $0 \leq t \leq t_0$, which satisfy the inequalities and minimize $y + z + t$. If the first inequality, say, is strict, $x + z < \eta(x + y)$, we can satisfy the inequalities with a smaller value of y. Similarly for the other two inequalities. Thus we assume that

$$x + z = \eta(x + y), \quad x + t = \eta(x + z), \quad y = \eta(x + t).$$

It is easy to see that these equalities imply $y > z > t$, and since $y \leq z + t$, also $t > 0$. Substituting $y = \eta(x + t)$ in the first equation yields $x + z = x(\eta + \eta^2) + \eta^2 t$, and substituting that in the second equation yields $x(\eta^3 + \eta^2 - 1) = t(1 - \eta^3)$. Thus one solution is $x = 1 - \eta^3$, $t = \eta^3 + \eta^2 - 1$, which implies $y = \eta^3$, $z = \eta^3 + \eta - 1$, and all other solutions are proportional to this one. The inequality $y \leq z + t$ now becomes $\eta^3 + \eta^2 + \eta - 2 \geq 0$, and the least η satisfying this is the unique positive root of $X^3 + X^2 + X - 2 = 0$. This root is $0.811 \cdots$. The other two inequalities $z \leq t + y$ and $t \leq y + z$ become $\eta^3 + \eta^2 - \eta \geq 0$ and $\eta^3 - \eta^2 + \eta \geq 0$, of which the first is implied by the previous inequality and the second holds for all $0 < \eta < 1$. Thus the previous values for the weights of the generators, with our particular η, yield an admissible weight on Γ, and for the rest of this section we assume that Γ is endowed with that weight. Note that choosing η to be a root of the polynomial above yields $w(d) = t = 1 - \eta$ and $w(c) = z = 1 - \eta^2$. The weight of any element $x \in \Gamma$ is a linear combination, with non-negative integer coefficients, of x, y, z, t. Here t is the smallest of the four numbers, therefore if $l(x) = k$, then $w(x) \geq kt$, and $k \leq \frac{w(x)}{t} \leq 6w(x)$. Fixing some weight w, the number of possible weights not exceeding w is then at most $(6w(x) + 1)^4$. In particular, there are only a finite number of possible weights below any given weight, and therefore it is allowed to apply induction by weights.

Theorem 10.24 *Let $\eta = 0.811 \cdots$ be the positive root of $X^3 + X^2 + X -$*

$2 = 0$, *and let* $\beta = \frac{\log 2}{\log 2 - \log \eta} = 0.767 \cdots$. *Then we can find a number* $B > 1$ *such that* $s_{\Gamma,w}(n) \leq B^{n^\beta}$.

Proof We will give two proofs. The first one employs a more or less direct attack, applying the weights that were constructed above, and yields a slightly weaker result than the one stated in the theorem: we will show that the theorem holds for any value of β that is bigger than the one stipulated in the theorem. Since for each such β we may get a different value of B, this does not establish the full theorem. To get that, we will introduce, following [Ba 98], a new method of counting the elements of Γ (the specific weights that we are using were also introduced in [Ba 98], without explanation). We proceed to the first proof.

First, note that we can find a B such that the required inequality holds for all elements up to some weight. Let $x \in \Gamma$. If $x \notin H$, then we can write $x = ya$ or $x = yu$, where $y \in H$, the element u is one of b, c, d, and $w(y) = w(x) - w(a)$ or $w(y) = w(x) - w(u)$. It follows that the number of elements of weight at most w is bounded by five times the number of elements of H of weight at most w. Write $\phi(y) = (y_l, y_r)$. Note that if $w(x)$ is large enough, the η inequality implies that y_l and y_r have smaller weights. We thus can apply induction to estimate the number of possibilities for y_l and y_r. The number of possibilities for y is then at most the product of these two numbers. If y_l and y_r have weights w_1 and w_2, then there are at most $(6w + 1)^4$ possibilities for w_1 and w_2. Thus we have, for n which need not be an integer,

$$s_{\Gamma,w}(n) \leq 5 \sum_{(w_1, w_2)} B^{w_1^\beta} B^{w_2^\beta}.$$

For $\beta < 1$ we have $w_1^\beta + w_2^\beta \leq 2(\frac{w_1 + w_2}{2})^\beta$, and thus the η inequality yields

$$s_{\Gamma,w}(n) \leq 5(6n + 1)^8 B^{2(\frac{\eta}{2}(n + 2w(a)))^\beta}.$$

We wish the right-hand side to be less than B^{n^β}. Taking logarithms, we have

$$\log 5 + 8 \log(6n + 1) + 2(\frac{\eta}{2}(n + 2w(a)))^\beta \log B \leq n^\beta \log B.$$

This inequality holds if $2(\frac{\eta}{2})^\beta < 1$, provided that n is large enough, and therefore it holds for all n, if we take a larger value of B. Thus we require that $\log 2 + \beta(\log \eta - \log 2) < 0$, i.e. that $\beta > \frac{\log 2}{\log 2 - \log \eta}$.   QED

An alternative proof of this result, with a generalization to an infinite class of groups, appears in [MP 01], but now we want to show that the result holds with the value of β that is given in the theorem. This seemingly small improvement requires a new method. It consists in associating to each element in Γ a tree, indeed a binary-rooted, labelled tree. The labels are the elements 1 and a. Given $x \in \Gamma$, there exists a unique element $y \in H$ such that $x = y$ or $x = ya$, and we label the root of the tree by either 1 or a, respectively. Then we write $\phi(y) = (y_l, y_r)$, and connect the root to the tree corresponding to y_l on the left and to the tree corresponding to y_r on the right. More precisely, we "grow", so to speak, the tree level by level. The root, with its label, is the first level. At the next level we have two vertices connected to the root, these vertices being considered as roots of the trees corresponding to y_l and y_r, labelled by 1 or a in the same way as the main root. We then connect these two roots to the roots of four further trees, determined by $\phi(y_l)$ and $\phi(y_r)$, etc.

The trees constructed in this way are infinite. For counting purposes, it is better to have finite ones. To that end we fix some positive number C, and let $x_1, \ldots, x_r$ be all the elements of weight at most C (we choose C so that $r > 0$). If $x = x_i$, for some i, the tree corresponding to x will contain only the root, labelled by the element x_i itself. If $w(x) > C$, we write $x = y$ or $x = ya$, as above, and label the root by 1 or a as before. If C is big enough, the η inequality implies that y_l and y_r have weights smaller that $w(x)$. We choose such a C, and then we may assume, inductively, that the trees corresponding to y_l and y_r are already known, and we connect these trees to the root on the left and right, respectively. This process associates to x a finite labelled tree, in which each vertex is connected to one vertex on the previous level (excepting the root), and to either two or no vertices on the next level. The vertices that are not connected to the next level are called *leaves*, and they are labelled by one of the elements $\{x_1, \ldots, x_r\}$, while the other vertices are labelled by either 1 or a. It turns out to be most convenient to count trees by the number of their leaves. First we consider unlabelled trees.

Lemma 10.25 *A binary-rooted finite tree with n leaves has $2n - 1$ vertices.*

The proof is immediate by induction.

Lemma 10.26 *The number of binary-rooted finite trees with n leaves is at most $\binom{2n}{n}$.*

Proof By induction, the case $n = 1$ being obvious. If $n > 1$, then the root is connected to two subtrees, with k and l leaves, say, where

$k+l = n$. Using induction, we obtain that the number of trees is at most $\sum_{k,l \geq 0,\ k+l=n} \binom{2k}{k}\binom{2l}{l}$. We will show that each summand is bounded by $\binom{2n}{n}/n$, which will prove the lemma. Thus, $\binom{2n}{n}$ is the number of ways to fill n positions out of given $2n$ ones. One of the ways to choose these positions is to choose k from the first $2k$ positions, and l from the rest. This gives $\binom{2k}{k}\binom{2l}{l}$ possibilities. We can take any of the k first filled positions and move it to one of the vacant positions among the later $2l$ positions. This shows that each of the initial choices gives rise to $kl \geq n - 1$ further choices, proving our claim.   QED

Lemma 10.27 $\binom{2n}{n} \leq 4^n$.

The proof is an easy induction.

Lemma 10.28 *The number of binary labelled rooted finite trees with n leaves is at most D^n, for some number D.*

Proof By the previous two lemmas, the number of unlabelled trees is at most 4^n, and by Lemma 10.25 the number of ways to label a given tree is $\frac{1}{2}(2r)^n$.   QED

We now return to the proof of the theorem (with the stipulated value of β). By the previous lemma, it will suffice to show that if $w(x) = w$, then the tree associated to x has at most Ew^β leaves, for some constant E. For technical reasons it is easier to establish the superficially stronger statement:

there exists a constant E, such that if $w(x)$ is large enough, then the tree associated to x has at most $E(w(x) - \eta w(a))^\beta$ leaves.

We assume, first of all, that $w(x) > C$. Writing as above $x = y$ or $x = ya$, and $\phi(y) = (y_l, y_r)$, we know that y_l and y_r have smaller weights than x. The number of leaves in the tree associated to x is the sum of the same numbers for y_l and y_r. If y_l and y_r have weights exceeding C, then, applying the η inequality and the facts $\eta < 1$ and $2(\eta/2)^\beta = 1$, the number of leaves for the x tree is at most

$$E(w(y_l) - \eta w(a))^\beta + E(w(y_r) - \eta w(a))^\beta$$
$$\leq E \cdot 2((w(y_l) + w(y_r) - 2\eta w(a))/2)^\beta$$
$$\leq E(2(\eta/2)^\beta(w(x) - w(a))^\beta \ < \ E(w(x) - \eta w(a))^\beta.$$

Let N be the maximal number of leaves in the trees associated to elements of weight smaller than C. To prove our claim by induction, we still have to establish it in the case that either y_l or y_r, or both, have weight smaller than C. We fix some number $\eta < \theta < 1$, and choose C so that for $w(x) > C$ we have $\eta(w(x) + w(a)) < \theta w(x)$. Let w_0 be the minimal weight which is bigger than C. We choose E big enough so that we have, first, $E(w_0 - \eta w(a))^\beta > 2N$, and, next, so that $E(\theta w(x) - \eta w(a))^\beta + N < E(w(x) - \eta w(a))^\beta$, whenever $w(x) > C$ (we leave it to the reader to verify that it is possible to choose E in that way). This choice of E provides us with the initial step of the induction and completes the proof.

As for lower bound, Grigorchuk's bound $\alpha = \frac{1}{2}$ (Corollary 10.22) was first improved by Y.G. Leonov [Le 00] to $0.506 \cdots$, and then by Bartholdi [Ba 01], to $0.5157 \cdots$

We state now without proof several further properties of Grigorchuk's group. Proofs can be found in [Hr 00].

(1) Γ is just infinite.

    This means that each proper factor group of Γ is finite. Indeed, it can be shown that each non-identity normal subgroup contains one of the subgroups H_n constructed in the proof of Theorem 10.12.

(2) Γ contains elements of arbitrarily high orders.

(3) Γ is not finitely presented (but it is recursively presented).

We also state some open problems; these are possibly very hard.

Problem 1 Does there exist a finitely presented group of intermediate growth?

Problem 2 Does there exist a group of intermediate growth with all elements of bounded orders?

The third problem does not mention growth, but we include it since we have discussed the Burnside problem.

Problem 3 Does there exist an infinite finitely presented torsion group?

11

More Groups of Intermediate Growth

Clearly, once we have constructed one group of intermediate growth, we can find many others, e.g. by taking direct products of our group either with itself or with groups of polynomial growth, or by taking finite extensions, etc. Many other constructions of groups of intermediate growth were offered as well. In this chapter we first describe a generalization, due to Grigorchuk himself, of the construction of the previous chapter. Then we will describe other approaches to the same groups; these approaches lead to many other interesting groups. Of these we supply some examples, but no proofs.

11.1 The General Grigorchuk Groups

We consider again transformations of the unit interval with the dyadic rationals removed. We divide that interval in the same way, and let E and P denote the same transformations, as before. We now let Γ be the group generated by four transformations a, b, c, d, where $a = P$ is the same as before, and each of b, c, d acts on the subintervals $(0, 1/2)$, $(1/2, 3/4), \ldots$ by some sequence of transformations P and E, e.g. P, P, P, E, E, P, E, E, E, $\ldots$. Each such sequence is allowed, but we assume that the three sequences defining b, c, d are related by requiring that on each subinterval two of b, c, d act as P, and the third one as E. Thus it is still true that the four generators have order two, and that any one of b, c, d is the product of the other two. It is convenient to let $0, 1, 2$ denote the three sequences $\{P, P, E\}$, $\{P, E, P\}$, and $\{E, P, P\}$ (think of them as columns), and then describe Γ by writing an infinite sequence λ of the symbols $0, 1, 2$, where the nth term of λ describes the action of b, c, d, respectively, on the nth subinterval. From now on we will describe our group as G_λ, reserving the letter Γ for the group of Chapter 10, which corresponds to the periodic sequence $0, 1, 2, 0, 1, 2, \ldots$, and which is usually called the *first Grigorchuk group* (we also append the index λ to the three generators b, c, d, omitting that index sometimes when there is no danger of confusion). Permuting the rows defining b, c, d changes only the names of the generators, but not the group, therefore we may, and will, assume that λ starts with 0. We let Λ be the set of all sequences, and we write it as $\Lambda = \Lambda_0 \cup \Lambda_1 \cup \Lambda_2$, where Λ_0 is the set of sequences in which each of $0, 1, 2$ occurs infinitely many times, in Λ_1 two of the symbols occur infinitely many times, and in Λ_2 only one symbol occurs infinitely often. For each $\lambda \in \Lambda$, we let $\sigma\lambda$ be the right shift of λ, i.e. the sequence λ with the first term removed (note that our convention that λ starts with 0 need not apply to $\sigma\lambda$).

Theorem 11.1

(a) *The groups Γ_λ are infinite and residually-2. If $\lambda \in \Lambda_0$, then Γ_λ is a 2-group, otherwise Γ_λ contains elements of infinite order, but all elements of finite order are 2-elements.*

(b) *If $\lambda \in \Lambda_2$, then Γ_λ is abelian-by-finite. Otherwise Γ_λ is of intermediate growth.*

Proof Let H_λ be the subgroup of index 2 in Γ_λ, which preserves the two halves of the unit interval. Then

$$H_\lambda = \langle b_\lambda, c_\lambda, d_\lambda, ab_\lambda a, ac_\lambda a, ad_\lambda a \rangle,$$

and writing for each element its action on the two halves, we obtain a 1–1 map

$$\phi_\lambda : H_\lambda \to \Gamma_{\sigma\lambda} \times \Gamma_{\sigma\lambda}.$$

Since λ starts with 0 (which equald PPE), both b_λ and c_λ induce a on $(0, 1/2)$, while $ab_\lambda a$ and $ac_\lambda a$ induce there $b_{\sigma\lambda}$ and $c_{\sigma\lambda}$. This means that $\phi_\lambda(H_\lambda)$ projects, as a subgroup of $\Gamma_{\sigma\lambda} \times \Gamma_{\sigma\lambda}$, onto the first factor; similarly, it projects also onto the second factor. Continuing, $\Gamma_{\sigma\lambda}$ has a subgroup of index 2 which projects onto $\Gamma_{\sigma^2\lambda}$, and so on. Since this series of projections goes on indefinitely, Γ_λ is infinite.

For each n, we divide $(0, 1)$ into 2^n equal subintervals. Γ_λ permutes these 2^n intervals, and each element of Γ_λ is determined by its action on these intervals for all n. This shows that Γ_λ is residually finite. We want to show that for each n, Γ_λ induces a 2-group on the 2^n intervals. This is clear for $n = 1$, and for bigger n we may assume, inductively, that

H_λ, which acts on each of the two subsets of size 2^{n-1} of subintervals of $(0, 1/2)$ and $(1/2, 1)$ like $G_{\sigma\lambda}$, acts on them as a 2-group. Since $|\Gamma_\lambda : H_\lambda| = 2$, this proves our claim and shows that Γ_λ is residually-2. If $x \in \Gamma_\lambda$ has odd order, it is mapped to the identity in any homomorphism onto a finite 2-group, and thus $x = 1$. Thus all elements are either 2-elements or have infinite order.

The generators of Γ_λ have order 2. We consider words of length 2. For example, $(ad_\lambda)^2 = ad_\lambda a \cdot d_\lambda$ is mapped by ϕ_λ onto $(d_{\sigma\lambda}, d_{\sigma\lambda})$, which has order 2, and thus ad_λ has order 4. On the other hand, $(ab_\lambda)^2$ is mapped onto $(b_{\sigma\lambda} a, ab_{\sigma\lambda})$. We square this element again, etc. If the row defining b_λ contains E somewhere, then eventually the repeated squaring will lead us to an element in some direct power of one of our groups with all components of length 1, hence of order 2, and then ab_λ is a 2-element. But if E does not occur anywhere, we see that no power $(ab_\lambda)^{2^k}$ is the identity, and ab_λ has infinite order. If $\lambda \in \Lambda_0$, we can employ induction as in the proof of Theorem 10.11 to show that Γ_λ is a 2-group. If $\lambda \in \Lambda_1 \cup \Lambda_2$, then we may assume that in the row defining b_λ the letter E occurs only finitely many times. Then for some k there are no occurrences of E in the row defining $b_{\sigma^k\lambda}$, therefore $\Gamma_{\sigma^k\lambda}$ contains elements of infinite order, and then so does Γ_λ, because it involves $\Gamma_{\sigma^k\lambda}$ as a section. Part **(a)** is now completely proved.

Next, let $\lambda \in \Lambda_2$. Then for some k the sequence $\mu := \sigma^k\lambda$ is constant. That means that among the three sequences defining b_μ, c_μ, d_μ, two are identical. If say, these are the two upper ones, then $b_\mu = c_\mu$, and thus G_μ is generated by the a, b_μ, and it is infinite dihedral, containing an infinite cyclic subgroup of index 2. The sequence of maps $\phi_{\sigma^i\lambda}$ shows that G_λ has a subgroup of finite index which is embedded in a direct power of G_μ, and therefore G_λ also contains an abelian subgroup of finite index.

Let $\lambda \in \Lambda_0 \cup \Lambda_1$. Since $\phi_\lambda(d_\lambda) = (1, d_{\sigma\lambda})$ and $\phi_\lambda(ad_\lambda a) = (d_{\sigma\lambda}, 1))$, we see that $\phi_\lambda(H_\lambda) \geq D \times D$, where D is the normal closure of $d_{\sigma\lambda}$. At least one of the two rows defining b_λ and c_λ contains E infinitely many times, therefore the calculation above shows that either $ab_{\sigma\lambda}$ or $ac_{\sigma\lambda}$ has a finite order. Since $G_{\sigma\lambda}/D$ is generated by the images of a and b, and also of a and c, we see that $G_{\sigma\lambda}/D$ is finite. It follows that $(G_{\sigma\lambda} \times G_{\sigma\lambda})/\phi_\lambda(H_\lambda)$ is finite, and since $H_\lambda \cong \phi_\lambda(H_\lambda)$, the groups G_λ and $G_{\sigma\lambda} \times G_{\sigma\lambda}$ are commensurable. Now suppose that G_λ has polynomial growth, say of degree n. Then Exercise 2.1 shows that $G_{\sigma\lambda}$ has growth of degree $\frac{n}{2}$, and for some k the group $G_{\sigma^k\lambda}$ has degree less than 1, which contradicts Example 1 of the Introduction. Thus G_λ does not have polynomial growth.

We will carry out the proof that G_λ has an intermediate growth only for $\lambda \in \Lambda_0$. This is simpler than the general case, and will furnish us with a large enough supply of groups of intermediate growth. The proof is similar to the one for Grigorchuk's first group Γ, but differs in details. First, if $x \in H_\lambda$, we write $\phi(x) = (x_l, x_r)$. If $l(x) = n$, then, as for Γ, we have $l(x_i) \le \frac{n+1}{2}$. Let $H_{\lambda,k}$ be the subgroup of G_λ for which ϕ^k is defined. Then for $x \in H_{\lambda,k}$ we can write $\phi^k(x) = (x_1, \ldots, x_{2^k})$, where $x_i \in G_{\sigma^k \lambda}$, and this time $l(x_i) \le \frac{n+2^k-1}{2^k}$. Note that in the last two inequalities, the lengths $l(x_i)$ are computed in the relevant groups, $G_{\sigma\lambda}$ and $G_{\sigma^k\lambda}$, respectively. We want to show that for an appropriate index k, actually for infinitely many indices, the sum $\sum_1^{2^k} l(x_i)$ is considerably less than the bound $n + 2^k - 1$ obtained from these inequalities.

To that end, we take some element $x \in H_{\lambda,k}$ of large length, write it as a word $w(a, b, c, d)$ of minimal length, and in w we note all occurrences of one letter, say b, and track it through all applications of ϕ^i. Since $x \in H_\lambda$, the word w can be separated onto products of the generators b, c, d, aba, aca, ada of H_λ. According to whether b occurs as itself or in aba, it is mapped by ϕ onto either $(a_{\sigma\lambda}, b_{\sigma\lambda})$ or $(b_{\sigma\lambda}, a_{\sigma\lambda})$. Thus each occurrence of b in x gives rise to one occurrence in $\phi(x)$, but it is possible that this occurrence is cancelled out. If in x occurs, e.g., the string $abadaca$, this string becomes $b \cdot 1 \cdot c = d$ in the first component of $\phi(x)$. Any such cancellation reduces the sum $l(x_l) + l(x_r)$ by one (at least). It is possible that such cancellations give rise to new occurrences of b in $\phi(x)$, as in our example we obtained a new occurrence of d, but we do not care about that: we keep track of only the original b occurrences. Since $\lambda \in \Lambda_0$, there is some index k such that $\lambda(k) = 2$. If some occurrence of b was not cancelled in the consecutive applications of ϕ before we arrived at the k place (by ϕ^{k-1}), at this place it is mapped onto either (E, b) or (b, E), and the occurrence of E shows that again $\sum l(x_i)$ is reduced by one. Thus if b occurs r times in x, then $\sum_1^{2^k} l(x_i) \le n + 2^k - 1 - r$. We need to estimate r.

Lemma 11.2 *If $l(x) \ge 5$, then a minimal word representing x does not contain the string dad.*

Proof Recall that $(ad)^4 = 1$. If dad occurs at the beginning or the middle of the word, then it is a part of a string $dadau$, where u is one of b, c, d. We can replace $dada = (da)^2$, of order 2, by its inverse $(ad)^2 = adad$, obtaining $adadu$, and du is $1, b$, or c, and thus $dadau$ can be replaced by a shorter string $adav$, which is a contradiction. This argument does not apply if $u = 1$, i.e. our word ends by *dada*, but then it ends by the string *adada*, which can be replaced by the shorter one $a(ad)^2 = dad$. Finally, if *dad* occurs at the end of the word, we have a string *uadad*, which can be replaced by *udada = vada*.   QED

Now in a minimal word representing x about half the letters are b, c, or d. The lemma shows that there are no two consecutive occurrences of d, therefore b and c occur at least $\frac{l(x)}{4}$ times, and one of them occurs at least $\frac{l(x)}{8}$ times. Since we overlooked the case of odd lengths, we compensate for that by saying that either b or c occurs at least $\frac{l(x)}{9}$ times. If b occurs that many times, it means that $r \geq \frac{l(x)}{9}$, and $\sum l(x_i) \leq (8/9)n + 2^k$. This is also true if c is the generator that occurs at least $\frac{l(x)}{9}$ times, if we replace k by l, the first index where $\lambda(l) = 1$.

We write $\omega(\lambda) = \omega(G_\lambda, \{a, b, c, d\})$. For each $\epsilon > 0$, there exists a number A such that $s_{G_\lambda, \{a,b,c,d\}}(n) \leq A(\omega(\lambda) + \epsilon)^n$. We use this notation for each λ. Write $l(x_i) = n_i$ and $s = \sum l(x_i)$. Then the numbers n_i are a partition of s, and they do not exceed $\frac{n+2^k-1}{2^k}$. The number of partitions is not more than $(\frac{n+2^k-1}{2^k})^{2^k}$. Given the partition, the number of possibilities for the sequence $\{x_i\}$ is at most $A^{2^k}(\omega(\sigma^k\lambda) + \epsilon)^s \leq A^{2^k}(\omega(\sigma^k\lambda) + \epsilon)^{(8/9)n+2^k}$. Thus the number of elements $x \in H_{\lambda,k}$ of length $\leq n$, in which b occurs at least $\frac{l(x)}{9}$ times, is at most $A^{2^k} \cdot (\frac{n+2^k-1}{2^k})^{2^k} \cdot (\omega(\sigma^k\lambda) + \epsilon)^{(8/9)n+2^k}$. The number of elements in which c is the generator occurring many times is given by a similar expression, with l replacing k. Taking nth roots, letting $n \to \infty$, and applying Lemma 10.21, shows that $\omega(\lambda) \leq (\omega + \epsilon)^{8/9}$, where $\omega = \max(\omega(\sigma^k(\lambda)), \omega(\sigma^l(\lambda)))$. Since both 2 and 1 occur infinitely many times in λ, we can repeat the process, to obtain that for each t there exists some ρ such that $\omega(\lambda) \leq \omega(\rho)^{(8/9)^t}$. But the numbers $\omega(\rho)$ are relative to a set of four generators, and therefore $\omega(\rho) \leq 7$ and $\omega(\lambda) = 1$.   QED

To obtain explicit bounds for the growth, we need to restrict our groups further.

Definition The sequence λ is *almost periodic* if there exists a number p, such that for each i, there exists $j \leq p$ such that $\lambda(i) = \lambda(i + j)$.

Theorem 11.3 *Let $\lambda \in \Lambda_0$ be almost periodic. Then there exist numbers $0 < \alpha < \beta < 1$, and $A, B > 0$, such that $A^{n^\alpha} \leq s_{G_\lambda}(n) \leq B^{n^\beta}$.*

Proof The proof of the upper bound is very similar to the corresponding proof of Theorem 10.16. We have to replace the group Γ there by either

G_λ or $G_{\sigma^p(\lambda)}$, depending on which side of the inequality we are, and replace the numbers 32 and 31 by 2^p and $2^p - 1$, and possibly change some more constants. The calculations yielding the lower bound are more involved, and we refer the reader to [Gri 84] for them, as well as for a proof that in many cases one can take $\alpha = 1/2$. Moreover, it follows from Theorem 13.6 below that the value $\alpha = 1/2$ is valid for all the groups G_λ with $\lambda \in \Lambda_0 \cup \Lambda_1$.   QED

We state, without proof, several more results from [Gri 84].

Theorem 11.4

(a) *Among the groups G_λ there are continuously many isomorphism types.*
(b) *There are continuously many mutually non-equivalent growth functions.*
(c) *There exist growth functions which are incomparable (up to equivalence).*
(d) *Given any function $f(n)$ which grows more slowly than any exponential function, there exist groups of intermediate growth whose growth is not dominated (up to equivalence) by $f(n)$.*

Theorem 11.5 *The groups G_λ, $\lambda \in \Lambda_0 \cup \Lambda_1$, are just infinite, not finitely presented, and the orders of their finite order elements are unbounded.*

In [Gri 85] Grigorchuk constructs p-groups, with odd p, and also torsion-free groups, with properties similar to those of the groups G_λ. He also points out that it follows that the finitely generated infinite p-groups, p odd, that were constructed earlier by V.I. Sushchanskii [Su 79], are of intermediate growth.

We are now going to describe, without proofs, several more constructions of groups of intermediate growth.

11.2 Groups Acting on Regular Trees

In chapter 10 we applied trees to estimate the growth of Grigorchuk's group. We now show how to use trees to construct this group and similar ones. Fix a number k, let $\mathcal{T} = \mathcal{T}_k$ be an infinite rooted k-regular tree, and let $\mathcal{A} = \mathcal{A}_k$ be the automorphism group of $\mathcal{T}$. This group fixes the root, and therefore fixes also, setwise, the nth level of the tree, which consists of the k^n vertices at distance n from the root. Since each automorphism is determined by its action on each level, the group $\mathcal{A}$ is residually finite. Consider the group G_n induced on the union of the first n levels. This has a normal subgroup N consisting of the automorphisms fixing each vertex at the $(n-1)$st level, and $N \cong S_k \times \cdots \times S_k$, with k factors (S_k is the symmetric group). It follows, by induction, that G_n is a Π_k-group, where Π_k consists of the primes up to k. The group $\mathcal{A}$ is the inverse limit of the $\{G_n\}$, and is thus a so-called *profinite group*, more precisely a pro-Π_k group. This is a huge group, of the cardinality of the continuum, but we are interested in its finitely generated subgroups.

Fixing some vertex $v \in \mathcal{T}$, all the vertices below it form a subtree $\mathcal{T}(v)$ that is naturally isomorphic to $\mathcal{T}$, with v as the root. We now take $k = 2$, and label the two vertices at level 1 as 1 and 2, from left to right. We denote by P the transformation that interchanges $\mathcal{T}(1)$ and $\mathcal{T}(2)$, according to their natural isomorphism, and E is the identity automorphism. Next, define four automorphisms a, b, c, d. The first one, a, is just P, while b, c, and d fix both 2 and 1, and thus they act on $\mathcal{T}(1)$ and $\mathcal{T}(2)$. Here b acts like a on $\mathcal{T}(1)$, and like c on $\mathcal{T}(2)$, while c also acts like a on $\mathcal{T}(2)$, but like d on $\mathcal{T}(2)$; and finally d acts like E (i.e. trivially) on $\mathcal{T}(1)$, and like b on $\mathcal{T}(2)$. This looks like a circular definition: we have defined b by means of c, defined c by d, and defined d by b! But this is not so: the definition is recursive. We know how b, c, d act on level 1. Assuming that we know already how they act on level n, we want to find out how they act on level $n + 1$. But the vertices of that level are the vertices of level n in $\mathcal{T}(1)$ and $\mathcal{T}(2)$, and since b, c, d fix these two subtrees, we know already how they act on their nth level, and thus we know the action on the $(n+1)$st level of $\mathcal{T}$. Let the two subtrees correspond to the two halves of the unit interval. Each subtree is itself divided in two parts, and we let the two halves of $\mathcal{T}(2)$ correspond to the two right quarters of the unit interval, etc. It is now clear that a, b, c, d generate a subgroup of $\mathcal{A}$ isomorphic to the first Grigorchuk group.

All the Grigorchuk groups discussed in the previous section can be realized as subgroups of $\mathcal{A}_2$, but for most of them we do not have recursive definitions, but have to specify their action on the subtrees, similarly to the action on subintervals that was specified in that section. We can also define similar groups acting on $\mathcal{T}_k$ for $k > 2$. The generators will again be denoted by a, b, c, d. Here $\mathcal{T}_k$ contains k isomorphic subtrees whose roots are the vertices at level 1; a permutes these subtrees cyclically, while b, c, d fix them setwise, act on the first one as some power of a, fix the next $k - 2$ elementwise, and act on the last one by dividing it again to k subtrees, and act on them in a similar manner. To specify b, c, d we have to specify the powers of a that act, first on the first subtree, then on the first "subsubtree" of the last subtree, etc., taking care to have $d = bc$. These specifications can be given by an infinite sequence on the letters $0, 1, \ldots, k-1$, and Grigorchuk shows that when $k = p$ is prime, if the sequence contains each of the p letters infinitely often, then the resulting group is a p-group of intermediate growth.

It is convenient to label all the vertices of $\mathcal{T}_k$ (except for the root). The ones at the first level are labelled as 1 to k. At the n level the label is a sequence of length n: for a vertex v we first write the label of the vertex u immediately above v, and then add one of the letters 1 to k, according to the position of v below u. Thus the vertices of $\mathcal{T}_k$ are labelled by all the finite sequences on the letters 1 to k, and we can identify each vertex with its label. The subtrees $\mathcal{T}(1), \ldots, \mathcal{T}(k)$ are the sets of sequences starting by $1, \ldots, k$, respectively, etc. With this identification, a vertex v is connected to the vertices vi, $i = 1, \ldots, k$. The isomorphism between $\mathcal{T}$ and $\mathcal{T}(v)$ is given by $u \to vu$. Given an automorphism α of $\mathcal{T}$, let $\alpha(v) = u$, then α maps $\mathcal{T}(v)$ to $\mathcal{T}(u)$, and there exists an automorphism β such that this action is described by $\alpha(vw) = u\beta(w)$. We write $\beta = \alpha_{|v}$, and call this automorphism the *restriction of α at v*. For the transformation a of the previous section, all restrictions are the identity, while for the generators b, c, d of Grigorchuk's first group Γ, the restrictions are either the identity or one of a, b, c, d. It follows that for each element of Γ, all of its restrictions are also in Γ. We formalize this situation.

Definition   A set S of automorphisms of $\mathcal{T} = \mathcal{T}_k$ is *self-similar*, if each restriction of an element of S lies in S.

The automorphisms of a self-similar set can be given by a recursive definition: we have to say how they act on the first level, and what are their restrictions to the vertices of that level. It is clear that there are only countably many finite self-similar sets, therefore most groups of the previous section cannot be generated by such sets.

11.3 Groups Defined by Finite Automata

In 1963 J. Horejs applied finite automata to define groups acting on the set of all words over some alphabet [Ho 63]. As we saw above, this is the same as defining groups acting on a regular tree. In 1972 S.V. Aleshin applied this construction to give a simple construction of an infinite finitely generated torsion group [Al 72], and in 1980 Grigorchuk published the construction of his first group, for the same end. Later he noticed that his group had intermediate growth, and Y.I. Merzlyakov noticed that Grigorchuk's and Aleshin's groups were commensurable, and hence had the same type of growth. We will now describe the construction of groups by finite automata.

First, a word of caution: there exists a very large class of groups termed *automatic* groups. This is a different notion from groups defined by automata, though these automata enter into their definition. For their theory, see, e.g., [ECHLPT 92] (another caution: their definition of a finite automaton differs from ours).

A *finite automaton of degree k* is a finite directed and labelled graph. Multiple edges between vertices, and loops, are allowed. From each vertex come out k edges, which are labelled by the numbers 1 to k, and each vertex is labelled by some permutation in S_k. To avoid confusion with the vertices of the trees T_k, we refer to the vertices of the automaton as *states*.

Given an automaton $\mathcal{M}$, we associate to each state s an element $\alpha_s \in \mathcal{A}_k$ as follows. We identify the vertices of T_k with their labels. Given a vertex $v = \{i_1, \ldots, i_n\} \in T_k$, we start at s and follow the edges labelled $\{i_1, \ldots, i_n\}$, visiting on the way the states $s = s_1, s_2, \ldots, s_n$, say. Let the permutation σ_r written at s_r map i_r to j_r. Then we map v to the vertex $\{j_1, \ldots, j_n\}$.

This map preserves levels. Moreover, if u is the vertex above v, then $u = \{i_1, \ldots, i_{n-1}\}$, and $\alpha_s(u) = \{j_1, \ldots, j_{n-1}\}$, and thus α_s preserves the tree structure. Finally, if $w = \{h_1, \ldots, h_n\}$ is another vertex, let r be the first index at which w differs from v. Then both w and v arrive at the state s_r, and σ_r maps i_r and h_r to different letters, therefore the images of w and v under α_s are different, and α_s is a permutation of T_k. The group generated by all the automorphisms α_s, for all states $s \in \mathcal{M}$, is the group $G_\mathcal{M}$ *defined* by $\mathcal{M}$.

Example 1 Let $\mathcal{M}$ consist of a single state, with k loops. Then $G_\mathcal{M}$ is the cyclic group generated by the permutation labelling the unique state (Figure 11.1: instead of drawing k loops, we draw just one, and label it by the numbers $1, \ldots, k$).

Example 2 Let $\mathcal{M}$ be a triangle ABC. From A go out two edges, to B and to C. From B both edges lead to C, and at C we have two loops.

Figure 11.1

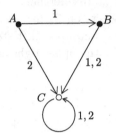

Figure 11.2

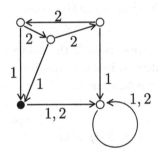

Figure 11.3

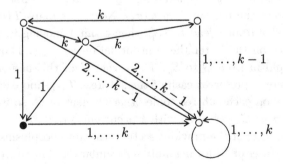

Figure 11.4

C is labelled by the identity, and A and B by the transposition $(1,2)$ (Figure 11.2: this diagram occurs already at [Al 72]).

Exercise 11.1 Check that $\mathcal{M}$ defines the dihedral group of order 8.

Example 3 Consider the automaton of Figure 11.3. Here $k = 2$, the bottom left state is labelled by the transposition $(1,2)$, and all other states by the identity. It is easy to see that the transformation determined by that bottom left state interchanges only the first element of each sequence v, and thus acts on T_2 like a. All other states preserve the two subtrees: the central and upper left states act on $T(1)$ like the bottom left, i.e. like a, while the upper right acts on it like the bottom right, i.e. trivially.

On $T(2)$ the upper left, central, and upper right states act like each other, cyclically permuted, and thus they act like b, c, d. Therefore that automaton defines the first Grigorchuk group.

Example 4 In Figure 11.4, the bottom left state is labelled by the cyclic permutation $(1, 2, \ldots, k)$, and the other states are labelled by the identity. This automaton defines the Grigorchuk group described at the third paragraph of the previous section.

The reader who has gone through the above examples will by now realize that a group G defined by an automaton can be given by a recursive definition of its action on some tree. If the automaton has degree k, then G acts on T_k: for each state s, the corresponding transformation α_s first permutes the subtrees $T(1), \ldots, T(k)$ according to the permutation $\sigma_s \in S_k$ labelling s, then, if $T(i)$ is moved to $T(j)$, then α_s induces on $T(j)$ the transformation α_t determined by the state t, which is the end point of the ith edge emanating from s. An element $x \in G$ can be written as a word in the generators, say $x = w(\alpha_s)$. First, x induces on the subtrees the permutation $w(\sigma_s)$. Next, if x maps $T(i)$ to $T(j)$, let the ith edge from s lead to $s(i)$, then x, after moving from $T(i)$ to $T(j)$, induces on the latter the transformation $w(\sigma_{s(i)})$. Thus we have a homomorphism from G to S_k. The kernel H of this homomorphism is of finite index, preserves each of the subtrees $T(i)$, and each element of H induces on each subtree some other element of G, so we have an embedding $H \to G \times \cdots \times G$, with k occurrences of G.

In the language of the previous section, the automorphisms of T_k defined by the states of $\mathcal{M}$ form a finite self-similar set. The converse is also true: any finite self-similar set S of automorphisms of T can be realized by an automaton $\mathcal{M}$, defined as follows: to each element $s \in S$ there corresponds some state of $\mathcal{M}$. This state is labelled by the permutation that s induces on the first level of $\mathcal{T}$, and if the restriction of s at i is t, then the ith edge leads from s to t.

Remark 1. A more general version of finite automata allows the labels of the states to be any map of the set $\{1, \ldots, k\}$ to itself, not necessarily a permutation. In that context our finite automata are termed *invertible*. The more general automata define semigroups, not groups, and some semigroups with interesting properties were constructed in this way, e.g. semigroups of polynomial growth of non-integral, and even irrational, degree.

Remark 2. In the literature, an automaton is often defined by giving the set of states, say A, a set X called the *alphabet* (in our definition, the numerals $1, \ldots, k$), and two functions from $A \times X$, one to A and one to X (in our notation, the first one takes (s, i) to the terminal state of the ith edge from s, and the second one takes (s, i) to $\alpha_s(i)$). Sometimes the two functions are combined into one, from $A \times X$ to itself (or to $X \times A$).

11.4 Bartholdi–Erschler Groups

In [BE 10] several remarkable constructions are described; they are summarized in the two theorems below. The first one defines the first groups of intermediate growth whose growth function is known (up to equivalence).

Theorem 11.6 *Let $\beta = 0.767 \cdots$ be the number occurring in Theorem 10.24. Then for each integer $r \geq 1$ there exists a 2-group K_r whose growth is equivalent to $e^{n^{1-(1-\beta)^r}}$ and a torsion-free group H_r with growth equivalent to $e^{\log n \cdot n^{1-(1-\beta)^r}} = n^{n^{1-(1-\beta)^r}}$.*

These groups are obtained by starting with $K_0 = C_2$ and defining recursively K_{r+1} as the wreath product of K_r with the first Grigorchuk group, relative to a certain transitive permutation representation of the latter. Starting with $\mathbb{Z}$, rather than C_2, yields groups whose growth is like the one stipulated for H_r. To find torsion-free examples, the authors start with H, a certain torsion-free group of intermediate growth that was also constructed by Grigorchuk, define $H_1 = H$, and H_{r+1} is a certain wreath product of H_r by H.

Theorem 11.7 *There exists a group G of intermediate growth containing a normal subgroup N, such that N is isomorphic to the group of all* *permutations of finite support of* $\mathbb{N}$, *and* $G/N \cong \Gamma$, *the first Grigorchuk group. The growth of* G *lies (up to equivalence) between the growths of* K_1 *and* H_1.

The exponent $1 - (1 - \beta)^r$ in Theorem 11.6 can be recovered from the growth function as $\lim \frac{\log \log s_n(G)}{\log n}$, and that limit, if it exists, can be considered as a rate of growth for groups of intermediate growth. In [Br 11] the following is proved:

Theorem 11.8 *For the above value of* β, *let* $\beta \le \gamma \le \delta \le 1$. *Then there exists a group* G *such that*

$$\liminf \frac{\log \log s_n(G)}{\log n} = \gamma \quad \text{and} \quad \limsup \frac{\log \log s_n(G)}{\log n} = \delta.$$

Here $\gamma = \delta$ ia allowed. This result shows the existence of continuously many incomparable (up to equivalence) intermediate growth functions, and also of continuously many comparable functions. The relevant groups are constructed as wreath products of finite groups by appropriate Grigorchuk groups G_λ.

12

Growth and Amenability

12.1 Amenability and Intermediate Growth

In this chapter and the next, we explore the relation of growth to other group theoretical properties.

Definition A group G is *amenable*, if it is possible to define on it a non-trivial finite, finitely additive, translation-invariant measure.

That means that we can find a function $\mu : 2^G \to \mathbb{R}^{\geq 0}$, which associates to each subset A of G a number $\mu(A) \geq 0$, and which satisfies:

(1) If $A, B \subseteq G$ and $A \cap B = \emptyset$, then $\mu(A \cup B) = \mu(A) + \mu(B)$.
(2) If $A \subseteq G$ and $x \in G$, then $\mu(Ax) = \mu(A)$.
(3) $\mu(G) > 0$.

The difference between this notion and the more customary notions of measure, such as Lebesgue or Haar measures, is that, first, we require μ to be defined for *all* subsets of G, and, on the other hand, we require additivity only for finite unions, not countable ones. From here on, whenever we say "measure", we usually mean one which satisfies properties (1)–(3) above.

Obviously multiplying μ by any positive constant yields another measure with the same properties, hence we will always assume that $\mu(G) = 1$. We required our measure to be invariant under right translations, but we can find also a left-invariant one, by defining $\nu(A) = \mu(A^{-1})$.

Example 1 Let G be finite. Then $\mu(A) = |A|/|G|$ is a measure, and clearly it is the only possible one.

On the other hand, if G is infinite, all finite subsets have zero measure.

Example 2 Consider the infinite cyclic group $\mathbb{Z}$. Recall the theory of limits relative to an ultrafilter, developed in Chapter 7. Fix some non-principal ultrafilter $\mathcal{F}$ on $\mathbb{N}$. For each subset A of $\mathbb{Z}$ define $A_n = \{x \in A \mid x \leq n\}$ and $\mu_n(A) = |A_n|/n$, and $\mu(A) = \mathcal{F} \lim \mu_n(A)$. Check that this is a measure on $\mathbb{Z}$.

This example shows that for infinite groups the measure is by no means unique.

Example 3 Let $F = \langle x, y \rangle$ be a free group of rank 2, with free generators x and y. Assume that μ is a (left-invariant) measure on F. Then $\mu(F - \{1\}) = 1$, and $F - \{1\} = A \cup B \cup C \cup D$, where A, B, C, D are the sets of reduced words starting by x, x^{-1}, y, y^{-1}, respectively. Then $A - \{x\} = xA \cup xC \cup xD$, a disjoint union. Since $\mu(A) = \mu(xA)$, it follows that $\mu(C) = \mu(D) = 0$. A similar argument shows that $\mu(A) = \mu(B) = 0$, and thus $\mu(F) = 0$, a contradiction. Thus F is not amenable.

On any space with a measure we can develop the notion of the integral defined by that measure. This is particularly simple in the case of an amenable group G, because all subsets of G are measurable, and thus all functions are measurable. Let $f(x)$ be a bounded real function on G, and let $a \leq f(x) < b$ (for all $x \in G$). Divide the interval $[a, b]$ in some way to subintervals, i.e. choose points $a = a_0 < a_1 < \cdots < a_k = b$, let Δ denote this subdivision, define $A_i = \{x \in G \mid a_{i-1} \leq f(x) < a_i\}$. The sets A_i constitute a partition of G. Put $S_\Delta = \sum_{i=1}^{k} \mu(A_i) a_i$. Then $S_\Delta \geq a \sum_{i=1}^{k} \mu(A_i) \geq a$. Similarly, if we put $s_\Delta = \sum_{i=1}^{k} \mu(A_i) a_{i-1}$, then $s_\Delta \leq b$. If E is a refinement of Δ, then $S_E \leq S_\Delta$ and $s_E \geq s_\Delta$. Any two divisions have a common refinement, therefore for any two divisions we have $s_E \leq S_\Delta$. Also $0 \leq S_\Delta - s_\Delta \leq max(a_i - a_{i-1})$. It follows that the infimum of the numbers S_Δ, taken over all subdivisions, equals the supremum of s_Δ. This common value is defined to be the integral of f relative to μ, denoted as usual by $\int f d\mu$.

It is clear that the integral is additive and right-invariant, in the sense that given any bounded function and an element $y \in G$, if we define $f_y(x) = f(xy)$, then $\int f_y d\mu = \int f d\mu$.

As a first application of integration we have:

Proposition 12.1 *If G is amenable, there exists on it a measure that is both right- and left-invariant.*

Proof Let μ be a measure on G, and define a new one by $\nu(A) =$

$\int \mu(xA)d\mu$. The verification that this is a two-sided invariant measure is immediate. QED

Definition Let $\mathcal{P}$ be any property of groups. A group G is *locally* $\mathcal{P}$ if each finitely generated subgroup of G has property $\mathcal{P}$.

Theorem 12.2

(i) *Finite and abelian groups are amenable.*

(ii) *Subgroups and factor groups of amenable groups are amenable.*

(iii) *An extension of an amenable group by an amenable group is amenable.*

(iv) *A locally amenable group is amenable.*

Proof (i). The claim about finite groups was already established. For abelian groups it follows from Examples **1** and **2** and the other parts of the theorem. Thus, the examples and part (iii) show that all finitely generated abelian groups are amenable, and now we invoke (iv).

(ii). Let G be amenable and $H \leq G$ and $N \triangleleft G$. A subset A of G/N is a collection $x_\alpha N$ of cosets of N, and we put $\mu(A) = \mu(\bigcup x_\alpha N)$. Next, let R be a set of representatives of the left cosets of H in G, and for $A \subseteq H$ put $\mu(A) = \mu(RA)$. This defines measures on G/N and H.

(iii). Let ν be a measure on N, and let σ be a left-invariant measure on G/N. For a coset $Nx \in G/N$ define $f(Nx) = \nu(N \cap Ax)$. This does not depend on the choice of the representative x, because if $y \in N$, then $\nu(N \cap Axy) = \nu(N \cap Ax)y = \nu(N \cap Ax)$, and thus f is a function on G/N. For $A \subseteq G$, put $\mu(A) = \int f d\sigma$. This is an additive function. Replacing A by Az $(z \in G)$ we change $f(Nx)$ to $f(Nzx) = f(zNx)$, implying $\mu(Az) = \mu(A)$ by the left invariance of the integral with respect to ν.

(iv). We first consider all functions from 2^G to the closed interval $[0, 1]$. This can be seen as the cartesian power $[0, 1]^{2^G}$, and with the product topology it is a compact space. The set of all finitely additive invariant measures μ, satisfying $\mu(G) = 1$, is defined by various equalities between values of μ, and therefore it is a closed set (possibly empty). Let H be a finitely generated subgroup of G, let μ be a measure on H, and extend it to G by setting $\mu(A) = \mu(A \cap H)$. The resulting measure on G is finitely additive and gives G measure 1, but it is invariant only with respect to multiplication by elements of H. The set $\mathcal{M}_H$ of all measures on G which are finitely additive and H-invariant is also closed, and we have just seen that it is not empty. Taking several finitely generated subgroups $H_1, \ldots, H_r$, and putting $K = \langle H_1, \ldots, H_r \rangle$, the sets M_{H_i} intersect in M_K, and therefore the intersection is not empty. Since the sets M_H are closed in a compact space, the intersection of all of them is not empty, and any function in that intersection is an invariant measure on G. QED

Exercise 12.1 Any group G contains a normal amenable subgroup which contains all normal amenable subgroups (this subgroup is termed the *amenable radical of G*).

The minimal class $\mathcal{E}$ of groups that satisfies items (i)–(iv) is termed the class of *elementary amenable groups*, or just elementary groups. Thus $\mathcal{E}$ is the smallest class of groups that contains all finite and all abelian groups, is closed with respect to forming subgroups, factor groups, and extensions, and it contains G whenever it contains all finitely generated subgroups of G (this last assumption, item (iv), is often replaced by the equivalent assumption that the relevant class is closed under direct limits). Theorem 12.2 states that all elementary groups are amenable. For many years it was not known if there existed other amenable groups. This will be answered soon, but first we need several more concepts and results. The first one is recalled from Chapter 2.

Definition Let the group G be generated by the set S, and let $A \subseteq G$. The *boundary ∂A* of A is the set of elements at distance 1 from A, i.e. the elements x such that $x \notin A$, but there exists an element $y \in A$ and a generator $s \in S$ such that $x = ys$ or $x = ys^{-1}$.

Definition A finitely generated group G satisfies the *Følner condition*, if $\inf |\partial X|/|X| = 0$, the infimum being taken over all finite subsets X of G. Equivalently, there exists in G a *Følner sequence*, i.e. a sequence X_n of finite subsets of G such that $\lim_{n\to\infty} \frac{|\partial F_n|}{|F_n|} = 0$.

Theorem 12.3 *A group satisfying the Følner condition is amenable.*

Proof Let X_n be a Følner sequence in G. As in Example 2, fix a non-principal ultrafilter $\mathcal{F}$, and for each subset A of G define $A_n = A \cap X_n$, $\mu_n(A) = |A_n|/|X_n|$, and $\mu(A) = \mathcal{F}\lim \mu_n(A)$. The additivity of μ is clear. Let $s \in S$ be one of the generators of G. Then $|As \cap X_n| = |A \cap X_n s^{-1}|$ and $A \cap X_n s^{-1} \subseteq A_n \cup \partial X_n$, and thus $|As \cap X_n|/|X_n| \leq \mu_n(A) + |\partial X_n|/|X_n|$, yielding, upon passage to the limit, $\mu(As) \leq \mu(A)$. In the same way we obtain $\mu(A) = \mu(As \cdot s^{-1}) \leq \mu(As)$, and thus $\mu(As) = \mu(A)$. QED

The converse of the theorem is also true; thus a group is amenable iff it satisfies the Følner conditions. We are not going to prove (or apply) that direction.

Corollary 12.4 *A group of subexponential growth is amenable.*

Proof Let G have subexponential growth, and let $B_n = \{x \in G \mid l(x) \leq n\}$. Then $|B_n| = s_n$, ∂B_n is the set of elements of length $n + 1$, and $\liminf s_{n+1}/s_n \leq \lim s_n^{1/n} = 1$. But $s_{n+1} \geq s_n$, and thus $\liminf s_{n+1}/s_n = 1$. But $s_{n+1} = |B_n| + |\partial B_n|$, and thus

$$\liminf |\partial B_n|/|B_n| = 0,$$

and there exists a subsequence of $\{B_n\}$ which is a Følner sequence.

QED

The following result is due to Ch. Chou [Ch 80]:

Theorem 12.5 *A finitely generated elementary amenable group either has exponential growth or is nilpotent-by-finite.*

Proof We first give a more explicit description of elementary groups. Let $\mathcal{E}_0$ be the class of finite or abelian groups, and define classes $\mathcal{E}_\kappa$, for each ordinal κ, as follows. If κ is a limit ordinal, put $\mathcal{E}_\kappa = \bigcup_{\lambda < \kappa} \mathcal{E}_\lambda$. If $\kappa = \lambda + 1$, let $\mathcal{E}_\kappa$ be the class of groups that are either locally in $\mathcal{E}_\lambda$, or are extensions of one group in $\mathcal{E}_\lambda$ by another. It is easy to see that each class $\mathcal{E}_\kappa$ is closed under taking subgroups and factor groups, and that $\mathcal{E} = \bigcup_\kappa \mathcal{E}_\kappa$, the union being taken over all ordinals.

Now let G be a finitely generated elementary amenable group, and let κ be the first ordinal such that $G \in \mathcal{E}_\kappa$. The theorem will be proved by transfinite induction on κ, the case $\kappa = 0$ being obvious. The minimality assumption shows that κ is not a limit ordinal, so write $\kappa = \lambda + 1$. If G is locally in $\mathcal{E}_\lambda$ then, since G itself is finitely generated, it lies in $\mathcal{E}_\lambda$, contradicting minimality. Thus there exists $N \triangleleft G$ such that N and G/N lie in $\mathcal{E}_\lambda$. Assume that G has subexponential growth. The induction hypothesis implies that G/N is nilpotent-by-finite, i.e. there exists a subgroup $H \triangleleft G$ such that $|G : H|$ is finite and H/N is nilpotent. Then H is finitely generated, and since H/N is polycyclic, repetition of the proof of Theorem 5.1 shows that N is finitely generated. By induction again, N is nilpotent-by-finite. Let K be a nilpotent subgroup of finite index in N. By Corollary 2.4, we may assume that K is characteristic in N, and then it is normal in G. Let $C = C_H(N/K)$, then H/C is a group of automorphisms of N/K, hence finite, and $|G/C|$ is also finite. Moreover $(C \cap N)/K = (C/K) \cap (N/K) = Z(N/K) \leq Z(C/K)$, and $C/C \cap N \cong CN/N \leq H/N$ is nilpotent, therefore C/K is also nilpotent, and C is soluble. We now can apply Corollary 5.4, the result on growth of soluble groups.

QED

Corollary 12.6 *The groups of intermediate growth are amenable but not elementary amenable.*

This is clear from the last two results. The Grigorchuk group Γ was the first example of an amenable group that was not elementarily amenable.

Once such an example was known, the following new problem arose: let S be the class of groups which is defined similarly to $\mathcal{E}$, but instead of requiring that it contains all finite or abelian groups, we require that it contains all groups of subexponential growth. Are all amenable groups in S? The answer is again negative [BV 05, Br 09].

Theorem 12.5 was strengthened by Osin [Os 04] to: *a finitely generated elementary amenable group of exponential growth has uniform exponential growth.*

Returning to Corollary 12.4 and its proof, there is one case in which we can point out explicitly a Følner sequence.

Proposition 12.7 *Let G be a finitely generated nilpotent-by-finite group. Then the sequence $\{B_n\}$ is a Følner sequence.*

This follows by combining Theorems 4.2 and 4.7. In the converse direction we have:

Proposition 12.8 *Let G be a finitely generated group for which $\{B_n\}$ is a Følner sequence. Then G has a subexponential growth.*

Indeed, the assumption means that $\lim s_{n+1}/s_n = 1$, and then also $\lim s_n^{1/n} = 1$.[1]

Amenability can be characterized by a notion akin to growth. Write a finitely generated group G, with d generators, in the form F/N, where F is a free group of rank d, and denote by $r_G(n)$, or r_n, the number of relations of G of length n, i.e. the number of elements of N of length n in F. Let $\rho(G) = \limsup r_n^{1/n}$. This number is the *cogrowth* of G.

Note that we cannot replace $\limsup$ in the definition by $\lim$. For example, consider a group G with defining relations of even length. Then all elements of N have even length (prove this!), and $r_n = 0$ for an odd n.

Proposition 12.9 *If $N \neq 1$, then $\sqrt{2d-1} \leq \rho(G) \leq 2d - 1$.*

---

[1] It is claimed in [Pi 00, Lemma 2.2] that in a group of exponential growth no subsequence of $\{B_n\}$ is a Følner sequence (this is also quoted, without proof, in [HR 00, VII.34]), but the proof there is wrong. I do not know if that claim, which would have been a nice characterization of subexponential growth, is true.

Proof The upper bound just states that the number of elements of length n in N does not exceed the number of elements of the same length in F. Let k be a number for which $r_k > 0$. Given an element of length k in N, say $r = x_1 \cdots x_k$, where each x_i is either a generator or its inverse, and a generator x (or x^{-1}), we obtain another element of N by looking at $x^{-1}rx$. This has length $k+2$, unless $x = x_1$ or x_k^{-1}. Therefore $r_{k+2} \geq r_k(2d-2)$. Next we apply the same process to the new relations, but since these relations now begin and end with a generator and its inverse, only conjugation by that one generator will not increase the length. Continuing, we obtain $r_{k+2n} \geq r_k(2d-2)(2d-1)^{n-1}$, implying $\rho(G) \geq \sqrt{2d-1}$. QED

We now quote the following, without proof.

Theorem 12.10 (Grigorchuk–Cohen) *The group G is amenable if and only if $\rho(G) = 2d - 1$.*

In other words, a group is amenable iff it has the maximal possible number of relations.

12.2 More Isoperimetric Inequalities

There is a natural connection between amenability and isoperimetry. First, if G is not amenable, then a strong isoperimetric inequality holds in G. Indeed, Theorem 12.3 shows that there exists some number $c > 0$, such that for all finite subsets $A \subseteq G$ the inequality $|A| \leq c|\partial A|$ holds. In the other direction, we will now apply amenability to prove the isoperimetric inequality that was alluded to in Chapter 2, just before Proposition 2.25.

We need some preparations. In the remarks following Proposition 2.25, we have defined a *geodesic* in a graph to be a path between two vertices, say x and y, whose length equals $d(x,y)$. Thus the length is finite. We now define an *infinite geodesic*, or a *line*, to be a sequence $\{x_n\}$, $n \in \mathbb{Z}$, such that $d(x_i, x_j) = |i - j|$. That implies in particular that x_n is connected to x_{n-1} and to x_{n+1}, but to no other vertices on the line. We will be interested in the lines together with the labelling of their vertices by the integers, as above, and to emphasize that we refer to them as *labelled lines*.

Proposition 12.11 *An infinite, locally finite, transitive and connected graph contains lines.*

Proof We noted in Section 2.2 that finite geodesics of any length exist. If α is a geodesic of length $2n+1$, it contains a unique point, its *midpoint*, at distance n from both ends of α. Fix some vertex x. By transitivity, there exist geodesics of length $2n + 1$ with x as their midpoint. Let L_n be the set of these geodesics. This is a finite set, and if $k < n$, there exists a map $L_n \to L_k$, mapping each geodesic to its unique segment which has length $2k + 1$ and x as its midpoint. The sets L_n with these maps form an inverse system of finite sets, therefore its inverse limit is non-empty. If $\{L_i\}$, $i \geq 0$ is a point of the inverse limit, then $\bigcup L_i$ is an infinite geodesic. Note that it can be labelled in such a way that $x = x_0$, and we term x_0 the *midpoint* of the line.   QED

We now consider a finitely generated group G, with an identity e, and its Cayley graph. In the proof of Proposition 2.26 we counted finite geodesics. Since the number of infinite geodesics through a given point is usually infinite, we have to replace the simple counting by arguments involving measure. We are not able to define a measure on the space of all (labelled) geodesics, but we will do so for a large enough subspace.

We start by choosing one infinite labelled line α, with e as its middle point. Our space, denoted by $\mathcal{T}$, consists of all left translations of α, i.e. all lines $x\alpha$, for all $x \in G$. Such a line has x as its midpoint. Let L be the set of all these lines which contain e. If $\beta \in L$, then $\beta = y\alpha$, for some y, and $e \in \beta$ means that $e = yx$, for some $x \in \alpha$. Thus $L = \{x^{-1}\alpha \mid x \in \alpha\}$. That sets up a 1-1 correspondence between L and α, which is in turn in 1-1 correspondence with $\mathbb{Z}$, and thus we have a 1-1 correspondence $x_n^{-1}\alpha \to n$ between L and $\mathbb{Z}$. Now choose some finitely additive invariant measure on $\mathbb{Z}$, and use our correspondence to define a measure μ on L, which we may assume to satisfy $\mu(L) = 1$. For any $x \in G$, the set xL consists of all lines in $\mathcal{T}$ which contain x, it is in 1-1 correspondence $\beta \to x\beta$ with L, and we use this correspondence to define a measure μ_x on xL. Given $A \subseteq L$ and $y \in G$, we have $yL = yx^{-1}(xL)$, $yA = yx^{-1}(xA)$, and $\mu_y(yA) = \mu A = \mu_x(xA)$.

Definition A subset $A \subseteq \mathcal{T}$ is *measurable*, if $A \subseteq \bigcup_1^n x_i L$, for some elements $x_1, \ldots, x_n \in G$.

The natural way to define measure for measurable sets is to associate to a set A as above a modified sum of the measures of its intersections with the sets $x_i L$, the modification needed to take care of overlaps. But A is a subset of more than one union $\bigcup_1^n x_i L$, and we have to show that our definition of measure is independent of the actual union chosen. We start with the simplest case.

Lemma 12.12 *Let $A \subseteq xL \cap yL$. Then $\mu_x(A) = \mu_y(A)$.*

Proof There are subsets $B, C \subseteq L$ such that $A = xB = yC$. Then $C = (y^{-1}x)B$, and multiplication by $y^{-1}x$ induces a 1–1 correspondence between B and C. If in the 1–1 correspondence between L and α a line $b \in B$ corresponds to a point $p \in \alpha$, then $b = p^{-1}\alpha$, and $y^{-1}xb = y^{-1}xp^{-1}\alpha$, so that $y^{-1}xb$ corresponds to $px^{-1}y$. Here $d(p, px^{-1}y) = l(x^{-1}y)$ is independent of p, therefore in the correspondence between L and $\mathbb{Z}$ the sets B and C correspond to sets which are translates of each other by $l(x^{-1}y)$, and have the same measure, implying that $\mu_x(A) = \mu(B) = \mu(C) = \mu_y(A)$. QED

Because of this lemma, we will now write simply $\mu(A)$ for $\mu_x(A)$ whenever $A \subseteq xL$, and we extend this notation for all measurable sets $A \subseteq \bigcup x_i L$ by

(M) $\qquad \mu(A) = \sum_{\emptyset \neq S \subseteq \{1,\dots,n\}} (-1)^{(|S|+1)} \mu(A \cap x_i L \cap x_j L \cap \cdots),$

where $S = \{i, j, \dots\}$.

In this equation μ, in the displayed term, can be interpreted as μ_{x_i}, or μ_{x_j}, and so on.

Proposition 12.13 *The equation (M) defines a well-defined, finitely additive, left-invariant measure on all measurable sets.*

Proof We have to prove that if $A \subseteq \bigcup x_i L$ and also $A \subseteq \bigcup y_j L$, then the two values obtained for $\mu(A)$ by (M) are equal. We may take the union of the two sets $\{x_1, \dots, x_n\}$ and $\{y_1, \dots, y_m\}$, obtaining one set $\{z_1, \dots, z_k\}$, and it suffices to prove that the x_i and z_j yield the same value. For that it suffices to show that if we add one set, say $x_{n+1}L$, on the right-hand side of (M), the value of the sum is not changed. But this addition adds to the right-hand side the sum

$$\sum_{S \subseteq \{1,\dots,n\}} (-1)^{|S|} \mu(A \cap x_{n+1}L \cap x_i L \cap x_j L \cap \cdots),$$

where $S = \{i, j, \dots\}$ (the empty set is included). In all the terms of this sum μ can be interpreted as $\mu_{x_{n+1}}$, and the finite additivity of μ on $x_{n+1}L$ and the exclusion–inclusion principle show that the sum vanishes. Thus the right-hand side of (M) is not changed.

The proofs of the additivity and invariance are easy and are left to the reader.   QED

We finally arrive at the isoperimetric inequalities.

Theorem 12.14 *Let G be a finitely generated infinite group. Then*

$$s_G(n) \le (n+1)a_G(n), \quad and \quad s_G(n) \le (n+1)a_G(n+1).$$

Proof We consider the Cayley graph Γ of G. Let $A \subseteq \Gamma$ be finite, with boundary ∂A. We consider the sets of the form $I = \beta \cap (A \cup \partial A)$, for some line β. In the rest of the proof, the letter I is reserved for sets of this form. Let L_I be the set of lines in $\mathcal{T}$ which intersect $A \cup \partial A$ in I. This is a measurable set, and $\bigcup_{x \in I} L_I = xL$. Since the sets L_I are disjoint, we have $\sum_{x \in I} \mu(L_I) = \mu(xL) = 1$. Thus

$$\sum_{I \subseteq A \cup \partial A} |I \cap A| \mu(L_I) = \sum_{I \subseteq A \cup \partial A} \sum_{x \in I \cap A} \mu(L_I) = \sum_{x \in A} \sum_{x \in I \subseteq A \cup \partial A} \mu(L_I)$$

$$= \sum_{x \in A} 1 = |A|.$$

Let $\beta \in L_I$, and assume that β intersects A. Considering some labelling of β, we see that among the points of $\beta \cap A$ there are first and last ones, and these are preceded and followed by two points of ∂A, say y, z, which lie in $\beta \cap \partial A \subseteq I$. Then

$$|I \cap A| = |\beta \cap A| \le d(y, z) \le d(y, e) + d(z, e) = l(y) + l(z)$$
$$\le \sum_{x \in I \cap \partial A} l(x).$$

This inequality certainly holds if $I \cap A = \emptyset$, and thus

$$|A| = \sum_{I} |I \cap A| \mu(L_I) \le \sum_{I} \left(\sum_{x \in I \cap \partial A} l(x) \right) \mu(L_I)$$
$$= \sum_{x \in \partial A} l(x) \sum_{x \in I} \mu(L_I) = \sum_{x \in \partial A} l(x).$$

Taking $A = B_n$, we obtain the second inequality of the theorem. For the first one, write $s_G(n-1) \le na_G(n)$, and add $a_G(n)$ to both sides.   QED

13

Intermediate Growth and Residual Finiteness

The Grigorchuk groups, and many other groups of intermediate growth, are residually finite. This need not be the case, as is obvious from Theorem 11.7. Below we will construct, with proofs, other examples, though these groups will still be close to being residually finite. In the other direction we will quote results that show that groups that are not only residually finite, but also residually nilpotent or residually soluble, and have slow subexponential growth, are actually of polynomial growth. We open this chapter with a result that points out that the connection between intermediate growth and residual finiteness may not be accidental. Note that groups of polynomial growth are residually finite, by Theorem 2.24.

Lemma 13.1 *A finitely generated group contains maximal normal subgroups.*

Proof Let G be finitely generated, let S be a finite set which generates G normally, i.e. $G = \langle S \rangle^G$, and choose S to have minimal possible cardinality. If T contains all elements of S but one, then $N := \langle T \rangle^G \neq G$ (T may be empty, in which case we take $N = 1$). By Zorn's Lemma, there exists a normal subgroup K which is maximal with respect to containing N and not containing S. Any normal subgroup containing K properly contains also S, and therefore equals G, thus K is a maximal normal subgroup. QED

Proposition 13.2 *A finite-by-nilpotent group is nilpotent-by-finite.*

Proof Let $N \triangleleft G$, with a finite N and a nilpotent G/N. Let $C = C_G(N)$. Then $C \triangleleft G$, and G/C is isomorphic to a subgroup of $\mathrm{Aut}(N)$, hence finite. Since $C \cap N = Z(N) \leq Z(C)$, and $C/C \cap N$ is nilpotent, C is also nilpotent. QED

Theorem 13.3 [BM 07] *Let G be a finitely generated group of intermediate growth. Let R be the intersection of all subgroups of finite index in G. Then either G/R is a residually finite group of intermediate growth, or R has a normal subgroup N such that R/N is a simple group of intermediate growth.*

Proof Clearly G/R is residually finite, so if it has intermediate growth, we are done. Since G/R, with G, has subexponential growth, there remains the case that G/R has polynomial growth. Then it is nilpotent-by-finite, and repeated application of Proposition 2.1 and Theorem 5.1 shows that R is finitely generated. Let N be a maximal normal subgroup of R. If R/N is finite, then N contains a finite index characteristic subgroup K, and then $K \lhd G$, and G/K is finite-by-nilpotent-by-finite, therefore nilpotent-by-finite and residually finite, contrary to the definition of R. Thus R/N is an infinite simple group. Since it has subexponential growth, and it is not nilpotent-by-finite, it is of intermediate growth. QED

Problem Does there exist a finitely generated simple group of intermediate growth?

For our construction of intermediate growth groups which are not residually finite, the following result, which we quote without proof, is basic.

Theorem 13.4 [Gri 99, Theorem 1] *There exists a finitely generated group G, such that $Z(G)$ is an infinite elementary abelian 2-group, and $G/Z(G) \cong \Gamma$, the first Grigorchuk group.*

Anna Erschler [Er 04] applied that theorem to prove that there are many groups of intermediate growth which are not residually finite.

Theorem 13.5

(a) *There exist $2^{\aleph_0}$ non-isomorphic groups of intermediate growth, which are not residually finite.*
(b) *There exists a group L of intermediate growth, which contains finite normal subgroups, but such that no factor group of L over a finite normal subgroup is residually finite.*

Proof We let G be the group of Theorem 13.4. Since G is finitely generated, it contains only countably many finite index subgroups. Let $\{N_1, N_2, \ldots\}$ be an enumeration of all finite index normal subgroup of G. The center $Z = Z(G)$ can be considered as a vector space of countable dimension over the field of two elements, and we apply in it additive notation and let $z_0, z_1, \ldots$ be a basis.

We want to construct two sequences $r(i)$, $s(i)$ of natural numbers, satisfying:

(i) $r(i)$ and $s(i)$ are positive even integers, all distinct from each other;

(ii) $z_{s(i)} N_i = z_{r(i)} N_i$.

Assuming that we have already found such integers for $i < k$, we let t be their maximum. Since G/N_k is finite, we can find two even numbers $r, s > t$, such that z_r and z_s have the same image in G/N_k, and we let these be $r(k)$ and $s(k)$. Thus the desired sequences are constructed.

Let C be the subgroup of Z generated by the elements $z_0 - z_{s(i)} + z_{r(i)}$. In Z/C, all the basis elements z_i with odd i remain independent, therefore Z/C is infinite. Also $z_0 \notin C$. Let $C \le K < Z$ be such that $z_0 \notin K$. We claim that G/K is not residually finite. To see that, note that if G/N is a finite image of G/K, then $N = N_i$ for some i, and since $(z_0 - z_{s(i)} + z_{r(i)}) \in C \le N$, we have $z_0 N = (z_{s(i)} - z_{r(i)}) N = N$ (see (ii)), i.e. $z_0 \in N$. Thus z_0 maps to the identity in any finite image of G/K, but not in G/K itself, because $z_0 \notin K$. That means that the intersection of all finite index subgroups in G/K contains a non-identity element, and G/K is not residually finite.

We can find a subgroup as above such that Z/K is finite, and then G/K is an extension of the finite group Z/K by Γ, and G/K has intermediate growth, by Proposition 2.5(c). We want to estimate the number of such Ks. Consider maps of Z onto C_2, in which z_0 is not mapped to the identity. We can specify the images of the basis elements $z_1, z_2, \ldots$ arbitrarily, therefore there are $2^{\aleph_0}$ such maps. Let H be any countable group. Since G is finitely generated, any homomorphism $G \to H$ is determined by the finitely many images of the generators of G, and there are at most countably many such maps, and they have at most countably many distinct kernels. Therefore among the $2^{\aleph_0}$ factor groups G/K there must be $2^{\aleph_0}$ non-isomorphic ones. That proves item (a).

In the groups that were just constructed we had $|Z/K| = 2$, and $G/Z \cong \Gamma$ is residually finite, thus these groups are not far removed from being residually finite. It is clear that the same method yields $2^{\aleph_0}$ groups in which Z/K is any finite elementary abelian 2-group, but we want to find also an appropriate K with Z/K infinite. To that end, partition the basis $\{z_i\}$ to infinitely many infinite subsets, obtaining accordingly a decomposition of Z as an infinite direct sum, $Z = Z_1 \oplus Z_2 \oplus \cdots Z_i \oplus \cdots$, where each Z_i is an infinite-dimensional subspace. Choose some basis element $x_i \in Z_i$. By repeating the previous construction, with x_i replacing z_0, and the elements $z_{s(i)}$ and $z_{r(i)}$ taken inside Z_i, we can find a subgroup $Y_i \leq Z_i$, such that $|Z_i : Y_i| = 2$, $x_i \notin Y_i$, and if $Y_i \leq K \leq Z$ and $x_i \notin K$, then G/K is not residually finite. Let $Y = Y_1 \bigoplus Y_2 \bigoplus \cdots$. Then Z/Y has the images of $\{x_i\}$ as a basis. Let $Y \leq X < Z$. Then some $x_i \notin X$, therefore G/X is not residually finite. In other words: not only is $H := G/Y$ not residually finite, it remains non residually finite on passage to any factor group over a proper subgroup of its centre (which is Z/Y, and is infinite). In particular, if N is a finite normal subgroup, then $NZ(H)/Z(H)$ is a finite normal subgroup of $H/Z(H) \cong \Gamma$, therefore $NZ(H)/Z(H) = 1$, i.e. $N \leq Z(H)$, and H/N is not residually finite. We want to show that we may choose X so that Z/X is infinite and G/X has intermediate growth.

Abusing notation, we let the letters x_i denote also the images of x_i in H. We write X_i for the subgroup of H generated by the x_k with $k \neq i$, and $H_i = H/X_i$. Then $x_i \neq 1$ in H_i (note that we have extended our abuse of notation to H_i). We measure length in H and H_i relative to generators which come from some fixed finite set of generators for G. Let $x \in H$ map onto $x_i \in H_i$. Then $xX_i = x_iX_i$, therefore $x \in Z/Y$. We can write $x = \sum_{j \in A} x_j$, for some finite set of indices A, and one of these indices must be i. Thus x can map only onto finitely many x_i, and the same applies to any finite set of elements of H, in particular to the elements of length at most n. That means that for i large enough, all x_i have length, in H_i, bigger than n: $\lim l_{H_i}(x_i) = \infty$.

We choose a subsequence y_i of x_i such that $l_{H_i}(y_i) > 2l_H(y_{i-1})$, let X be the subgroup generated by the other x_i, let T_i be the subgroup generated by all $x_k \neq y_i$ (this is one of our previous X_i), $L = H/X$, and $L_i = H/T_i$. Since we have maps $G \to H \to L \to T_i$, we have $l_L(y_i) \geq l_{T_i}(y_i) \geq 2l_H(y_{i-1} \geq 2l_L(y_{i-1}))$, and thus $l_L(y_i) \geq 2^{i-1}$. Let $y = \sum_{j \in A} y_j \in Z(L)$ (for some finite set A) have length at most n. Then $l_L(y) \geq l_{T_j}(y_j) \geq 2^{j-1}$, and thus $j - 1 \leq \log_2 n$, implying $A \subseteq \{1, 2, \ldots, \log_2 n + 1\}$, and the number of possibilities for A and y is at most $2^{\log_2 n + 1} = 2n$.

Now consider any element $x \in L$ of length at most n. Let u have minimal length in the coset $xZ(L)$. Then $l(u) \leq n$, and $x = uy$, for some $y \in Z(L)$, and $l(y) \leq 2n$. Therefore the number of possible ys is at most $4n$, while the number of possible us is at most $s_{L/Z(L)}(n)$, i.e. $s_L(n) \leq 4ns_{L/Z(L)}(n)$. Since $L/Z(L) \cong \Gamma$ has intermediate growth, so does L.   QED

We quote without proof two results about residually nilpotent and residually soluble groups.

Theorem 13.6 (Grigorchuk) *Let G be a finitely generated, residually nilpotent group, whose growth is strictly less (up to equivalence) than $2^{\sqrt{n}}$. Then G has polynomial growth.*

A proof is given in [DDMS 99], Interlude E. Note that a finitely generated residually nilpotent group is residually finite, by 2.24, and so are residually soluble groups of subexponential growth, by the same result combined with 5.2.

Theorem 13.7 (Wilson [Wi 11]) *Let G be a finitely generated, residually soluble group whose growth is strictly less (up to equivalence) than $2^{n^{1/6}}$. Then G has polynomial growth.*

14
Explicit Calculations

As already noted, for most groups explicit calculation of the growth function is very difficult, if not outright impossible. It is still remarkable that such calculations were carried out for many interesting groups. Simple examples were given already in the introductory chapter. Here we will give more examples. We also saw that for direct and free products, there are simple formulas for the growth, in terms of the growth of the factors. In this chapter we will derive several more formulas of that type. We start with direct calculations.

14.1 The Trefoil Group

The group $G = \langle x, y \mid x^2 = y^3 \rangle$ is known as the *trefoil group*, being the group of the trefoil knot. It is an amalgamated free product of the two infinite cyclic subgroups generated by x and y, amalgamating the subgroup of index 2 in the first with the one of index 3 in the second. It has further interest in being a one-relator group with a particularly simple relator. We will add to the generators of G the element $z = x^2 = y^3$, which is central, and calculate the growth function relative to this set of three generators.

We give first a short argument to show that $\omega(G, \{x, y, z\}) = \sqrt{2}$. Let $t \in G$ have length n and be represented by a word $w(x, y, z)$ of minimal length. Since z is central, we may assume that $w(x, y, z) = z^i u(x, y)$, for some i and some word u in x and y. Since $x^2 = y^3 = z$, we can replace each power x^e or y^e, with $|e| > 1$, by a word of the form $z^k x^{\pm 1}$ or $z^k y^{\pm 1}$, which is not longer. Therefore we assume that u contains x and y only to the power ± 1, and obviously the powers of x and y alternate. Now we replace x^{-1} by $z^{-1}x$. This changes $z^i u$ to $z^j v$, where the new exponent j is obtained by subtracting from i the number of occurrences of x^{-1} in u. We may have $|j| > |i|$, but we still have $|j| \leq n$,

$l(v) = l(u) \leq n$, and in v factors x alternate with factors $y^{\pm 1}$. The number of possibilities for $z^j v$ is then at most $(2n + 1)2^{(n+3)/2}$, yielding $\omega(G, \{x, y, z\}) \leq \sqrt{2}$. Since clearly $\omega(G, \{x, y\}) \leq \omega(G, \{x, y, z\})$, we obtain $\omega(G, \{x, y\}) \leq \sqrt{2}$. Adding the relation $x^2 = 1$ leads to the free product $H = C_2 * C_3$, and thus there is a homomorphism from G onto H, which shows that $\omega(G, \{x, y\}) \geq \omega(H, \{x, y\}) = \sqrt{2}$ (see Example 10 of Chapter 1 on page 7). Thus $\omega(G, \{x, y, z\}) = \omega(G, \{x, y\}) = \sqrt{2}$. It can be shown that $\Omega(H) = \sqrt{2}$, therefore also $\Omega(G) = \sqrt{2}$ (see the comments following Theorem 16.12 below).

A similar argument shows that $\Omega(\langle x, y \mid x^2 = y^4 \rangle) = \Omega(C_2 * C_4) = \zeta$, where ζ is the famous *golden ratio* $\frac{1+\sqrt{5}}{2}$ (see the same comments).

We pass to the generating growth function. As above, we represent each element $t \in G$ by a word $w = z^i u(x, y)$, where u contains x and y only to the power ± 1, and the occurrences of x and y alternate. Suppose that $i > 0$ and in u the letter x^{-1} occurs. Then we can replace x^{-1} by $z^{-1}x$, and thus replace i by $i - 1$, and replace u by a word of the same length, and the length of w is reduced. Similarly if $i < 0$ and x occurs in u. Therefore if $i \neq 0$, then in all occurrences of $x^{\pm 1}$ the exponent has the same sign, which is the sign of i.

Let $i = 0$. Suppose that u involves both x and x^{-1}. Then we can write $x^{-1} = z^{-1}x$, then move z^{-1} to the position of x, and replace there $z^{-1}x$ by x^{-1}. That means that we can interchange the occurrences of x and of x^{-1}, therefore we may assume that all terms x precede all terms x^{-1}. We assume that the minimal words with $i = 0$ are of that form, the xs preceding the x^{-1}s.

We now have all elements of G represented by words of the form $z^i u(x, y)$, where in u letters $x^{\pm 1}$ alternate with letters $y^{\pm 1}$, and either $i \neq 0$ and all occurrences of $x^{\pm 1}$ have the same exponent as the sign of i, or $i = 0$ and in u all occurrences x precede those of x^{-1}.

We make one more change. We replace all occurrences of x^{-1} by $z^{-1}x$. This move, which may increase the length, results in a word $w = z^i u(x, y)$ in which in u factors x and $y^{\pm 1}$ alternate. Note that these words u look like the elements of H, and they are in 1-1 correspondence with these elements. We claim that this form of the elements of G is unique.

Indeed, suppose that $z^i u(x, y) = z^j v(x, y)$, with u, v of the above form. Mapping G to H, we find the equality $u(x, y) = v(x, y)$ in H, and since H is a free product, that means that u and v are the same word, and the equality $u(x, y) = v(x, y)$ holds already in G. This in turn implies $i = j$.

Because of the uniqueness, we will refer to that form as the *canonical form* for the elements of G.

Now return to the minimal form. Assume first that $t = z^i u(x, y)$ with $i > 0$. If $s = t^j v(x, y)$ also has $j > 0$, then t, s are already in their canonical forms, and can be equal only if they are identical. Let $j \leq 0$, and let s involve $k > 0$ occurrences of x^{-1} (if $k = 0$ then necessarily $j = 0$ and s is already in canonical form). Then s has the canonical form $z^{j-k} v'(x, y)$, and since $j - k < 0$, we cannot have $t = s$. Thus the minimal form of t is unique, and, moreover, an element in the canonical form $z^i u(x, y)$, with $i > 0$, is already in its minimal form. The same applies to elements $z^j v(x, y)$ with $j < 0$ and v involving only x^{-1}, because these elements are just the inverses of the elements of the previous types.

There remains the possibility of an equality $u(x, y) = v(x, y)$, where in both u and v the xs precede the x^{-1}s, but this is also ruled out by going to the canonical forms. Thus the minimal forms are also unique, and now we can count them.

First, if $i = 0$, then the representing word is determined by giving k, the number of occurrences of x^{-1}, the initial letter ($x^{\pm 1}$ or $y^{\pm 1}$), and the distribution of the exponents ± 1 of the letters $y^{\pm 1}$. If $n = 2r$ is even, then k varies between 0 and r. If the initial position is $x^{\pm 1}$, then it is x if $k < r$ and x^{-1} if $k = r$, and if the initial position is $y^{\pm 1}$, then whether it is y or y^{-1} is determined by the distribution of the exponents, and thus we only have to know if the initial position is a power of x or of y – whether it is the element itself or its inverse is determined by the other data. Therefore we count two possibilities for the initial position, and 2^r possibilities for the distribution of $y^{\pm 1}$, and thus $(r + 1)2^{r+1}$ possibilities in all. If $n = 2r + 1$, a similar calculation shows that we have $(r + 2)2^r + (r + 1)2^{r+1} = 3r2^r + 2^{r+2}$ possibilities.

If to the minimal words with $i \neq 0$ we add the ones with $i = 0$, and only x and $y^{\pm 1}$ occurring, the number of words of a given length is the same as in the direct product $K = \mathbb{Z} \times (C_2 * C_3)$. Therefore, the generating growth function A_G of G is $A_K - A_{C_2 * C_3} + B$, where B is the generating function for the number of words with $i = 0$, which we just calculated. Here $A_{C_2} = 1 + X$ and $A_{C_3} = 1 + 2X$, implying that

$$A_{C_2 * C_3} = \frac{(1 + X)(1 + 2X)}{1 - 2X^2},$$

and

$$A_K - A_{C_2 * C_3} = A_{C_2 * C_3}(A_{\mathbb{Z}} - 1) = \frac{2X(1 + X)(1 + 2X)}{(1 - X)(1 - 2X^2)}.$$

Here we used $A_{\mathbb{Z}} = \frac{1+X}{1-X}$, and note that B is the sum of two infinite series. The first is

$$1 + \sum_1^\infty (r+1)2^{r+1}X^{2r} = 1 + 2\sum_1^\infty (r+1)(2X^2)^r = 1 + 2\sum_2^\infty r(2X^2)^{r-1}$$

$$= 1 + 2(\frac{1}{(1-2X^2)^2} - 1) = \frac{1+4X^2-4X^4}{(1-2X^2)^2}.$$

The second sum is

$$\sum_0^\infty (3r2^r + 2^{r+2})X^{2r+1} = \frac{6X^3}{(1-2X^2)^2} + \frac{4X}{1-2X^2} = \frac{4X-2X^3}{(1-2X^2)^2}.$$

Thus $B = \frac{1+4X+4X^2-2X^3-4X^4}{(1-2X^2)^2}$. Finally, A_G turns out to be the rational function

$$\frac{2X(1+X)(1+2X)(1-2X^2) + (1-X)(1+4X+4X^2-2X^3-4X^4)}{(1-X)(1-2X^2)^2}$$

$$= \frac{2X(1+X)(1+2X)(1-2X^2) + (1-X)(1+2X)(1+2X-2X^3)}{(1-X)(1-2X^2)^2}$$

$$= \frac{(1+2X)(1+3X-6X^3-2X^4)}{(1-X)(1-2X^2)^2}.$$

We now can deduce again $\omega(G, \{x, y, z\}|) = \sqrt{2}$, either from an explicit formula for $a(n)$, which can be derived from the above calculation, or by noting that $1/\sqrt{2}$ is the least absolute value of a pole of that rational function, and thus it equals the radius of convergence of the growth-generating function.

A different calculation, applying a more geometric language, of the same growth function, was given by M. Shapiro [Sh 94]. He also calculates the growth function relative to $\{x, y\}$, which is also rational, of a slightly more complicated form (that function was also determined in several other papers, see [JKS 95], [Gi 99]; the latter reference has a misprint in the formula).

14.2 Wreath Products

In this section we follow [Jo 91]. Recall that given groups K and H, their *standard wreath product* $K wr H$ is the split extension of B by H, where B, the *base group*, is the direct product of $|H|$ copies K_h of K, indexed by the elements of H, and H acts on B by moving the factors according to the multiplication in H, an element $x \in H$ moving K_h to K_{hx}. More formally, B is the set of functions from H to K with finite support, and if $f \in B$, $x, h \in H$, then $f^x(h) = f(h^{x^{-1}})$. We identify K with K_1. All other factors are then conjugate to K by some element of H, $K_h = K^h$.

If S and T are generating sets for K and H, then $S \cup T$ is a generating set for $G = K \,wr\, H$. Each non-identity element $t \in G$ can be written in the form $t = (\Pi_1^n x_i^{y_i})z$, with $x_i \in K$, $y_i, z \in H$, and some n, where if $t \in H$ then the product is empty, $n = 0$. Expressing x_i, y_i, z in terms of the generators, we have a word in $S \cup T$ of length $\sum l(x_i) + l(y_1) + \sum l(y_i y_{i+1}^{-1}) + l(y_n z)$. Write $Y = \{y_1, \ldots, y_n\}$, and given an element z and a finite subset Y of H (possibly empty), order Y in such a way that $l(y_1) + \sum_{i=1}^{n-1} l(y_i y_{i+1}^{-1}) + l(y_n z)$ is minimal. If there are several possibilities to order Y for that end, choose one of them. With that choice of the ordering of Y, the expression of t above is unique; it yields the minimal length among similar expressions, and will be termed the *canonical form*. We want to show that it is indeed a minimal form.

Let $t = w(s, t)$ be a minimal form for t, where s, t vary over S and T, respectively. Grouping together generators from S and T, we can write $w(s, t) = h_1 k_1 h_2 \ldots k_n h_{n+1}$, where $k_i \in K$, $h_i \in H$; here $n = 0$ is possible, and only h_1 and h_{n+1} can equal 1. Define $a_i = h_1 \cdots h_i$, and suppose that for some $i \neq j$ we have $a_i = a_j$. We may then assume that $i < j$, and that j is the first index following i for which the equality holds. Write $b_r = h_{i+1} \cdots h_r$, $c_r = b_r^{-1}$. Then $b_j = 1$, equivalently $h_j = c_{j-1}$, and $b_r \neq 1$ for $r < j$. The segment $d := h_{i+1} \cdots k_{j-1} h_j = k_{i+1}^{c_{i+1}} \cdots k_{j-1}^{c_{j-1}}$ is a product of conjugates of elements of K by non-identity elements of H, therefore d commutes with k_j, and in w we can interchange these two elements, bringing together k_i and k_j. This does not increase the length of w, but reduces the number of occurrences of elements of K, therefore after finitely many such changes we may assume that $a_i \neq a_j$ whenever $i \neq j$. With that assumption, and $e_i = a_i^{-1}$, we have $t = k_1^{e_1} k_2^{e_2} \cdots k_n^{e_n} e_n^{-1} h_{n+1}$, with $h_i = e_{i-1} e_i^{-1}$, and $l(t) = \sum l(k_i) + \sum l(h_i)$. Since the e_i are all different, this expresses t in the standard form of a wreath product element. Changing the order of the factors can change the value of $\sum l(h_i)$, but not of $\sum l(k_i)$, and if we now write y_i for e_i and z for $e_n^{-1} h_{n+1}$, we see that t is either in the canonical form, or in a similar form with the same length.

We now have to count the number of canonical forms. For any pair (Y, z), where $Y \subseteq H$ is finite and $z \in H$, let m be the corresponding minimal length. To determine an element t in canonical form associated to this pair we have to specify the elements x_i, non-identity elements of K. If $\sum l(x_i) = n$, then $l(t) = m + n$, the number of choices for the x_i is the number of elements of length n, with no trivial component, in the direct product of $|Y|$ copies of K. If the generating growth function of K is $C(X)$, this number is the coefficient of X^n in $(C(X) - 1)^{|Y|}$. Let $A(X)$ be the generating growth function for G. Since these x_i determine an element of length $m + n$ in G, we obtain $A(X) = \sum_{Y,z} u_m X^m (C(X) - 1)^{|Y|}$, where u_m is the number of pairs (Y, z) yielding the value m. These numbers u_m depend only on H, not on K. They were determined in some simple cases, see [Jo 91, Wo 97], and on principle they can be determined at least for any finite H.

Another possibility for the case that H is finite is to take as generators for B copies of S, one for each factor K_h, and then add T to obtain a generating set for G. Then in the semidirect decomposition $G = BH$, conjugation by elements of H preserves the generators of B, and by the results of Chapter 1 the growth-generating function with respect to that basis is $B(X)^{|H|} C(X)$, with $B(X)$, $C(X)$ being the corresponding functions for K and H. This much simpler method has the drawback of allowing a lot of superfluous generators.

14.3 Free Products with Amalgamations and HNN-Extensions

In the proofs of this section we often apply arguments similar to the proofs of Propositions 1.4 and 1.5. To avoid repetitions, we formalize these arguments. Given a finite alphabet $\mathbb{A}$, the set of all finite strings (*words*) formed from the letters of $\mathbb{A}$, including the empty string, is denoted by $\mathbb{A}^*$. It is the free monoid generated by $\mathbb{A}$. A *language* $\mathbb{L}$ over $\mathbb{A}$ is the set of equivalence classes for some equivalence relation on $\mathbb{A}^*$, e.g. the set of elements of a group G generated by $\mathbb{A}$ (which is supposed to include the inverses of its elements). Equivalently, we can choose a set of representatives for the equivalence classes and call this set a *language*. In this approach, which we will adopt from now on, any subset of $\mathbb{A}^*$ is a language. We let a_n be the number of words of length n in $\mathbb{L}$, and $A(X) = \sum a_n X^n$ is the *growth-generating function* of $\mathbb{L}$.

Now consider two disjoint alphabets $\mathbb{A}$ and $\mathbb{B}$, and languages $\mathbb{L}$ and $\mathbb{M}$ on them. We define two new languages on $\mathbb{A} \cup \mathbb{B}$. The first is the *direct product*, consisting of all words lm, with $l \in \mathbb{L}$, $m \in \mathbb{M}$, and the other is the *free product*, consisting of all strings $l_1 m_1 l_2 \cdots m_n$, with words from $\mathbb{L}$ and $\mathbb{M}$ alternating, and l_1 and m_n, but not the other words, may be the empty word.

Proposition 14.1 *In the situation above, assume that both $\mathbb{L}$ and $\mathbb{M}$ include the empty word, let $A(X)$, $B(X)$ be the generating growth functions of $\mathbb{L}$ and $\mathbb{M}$, and let $C(X)$, $D(X)$ be the corresponding functions for the direct and free products, respectively. Then*

$$C(X) = A(X)B(X),$$

$$\frac{1}{D} - 1 = \left(\frac{1}{A} - 1 \right) + \left(\frac{1}{B} - 1 \right).$$

The proof is identical to that of Propositions 1.4(b) and 1.5.

Definition   Consider pairs (G, S), where G is generated by S. A pair (H, T) is *admissible* in (G, S), if H is a subgroup of G, $T \subseteq S$, and there exists a transversal U for H in G, termed an *admissible transversal*, such that if $g = hu$, with $g \in G$, $h \in H$, $u \in U$, then $l_X(g) = l_Y(h) + l_X(u)$. We always assume that the transversal contains the identity as the representative of H.

We sometimes just say that H is admissible in G, suppressing the generating sets. The term *compatible*, rather than admissible, is used in [Le 91].

Exercise 14.1 If (H, T) is admissible in (G, S), and (K, R) is admissible in (H, T), then (K, R) is admissible in (G, S).

Proposition 14.2 ([Al 91, Le 91] *Let (L, R) be admissible in both (H, S) and (K, T). Let $G = H *_L K$. Let G, H, K, L have generating growth functions $A(X)$, $B(X)$, $C(X)$, and $D(X)$, respectively, relative to the generating sets $S \cup T$, S, T, R. Then*

$$\frac{1}{A} = \frac{1}{B} + \frac{1}{C} - \frac{1}{D}.$$

Proof   Let U, V be admissible transversals for L in H and K, respectively. As is well known, each element $x \in G$ can be written uniquely in the canonical form $x = l u_1 v_1 \cdots u_k v_k$, with $l \in L$, $u_i \in U$, $v_i \in V$ (here u_1, v_k, or the whole segment $u_1 \cdots v_k$ may be absent). We want to show that this form is a minimal one for x: more precisely, it becomes a minimal form after replacing all factors by minimal forms for them. Suppose, then, that $x = w(s, t)$ is a minimal form. Collecting together generators from the same group, we obtain a product in which elements from H and K alternate. If, say, $h \in H$ is one of these factors, we write it as $h = lu$. Replacing h by this product cannot increase the length of w, and performing this change on all the factors, we get a product in which elements from L alternate with elements from U and V. If we have in this product a segment of the form vl, say, then $vl \in K$, and we can replace it by an equal product of the form $l'v'$, again without increasing the length. In this way we 'collect to the left' all occurrences of elements of L, ending with a canonical form for x that is also minimal.

We let $E(X)$, $F(X)$ denote the generating growth functions for U and V (languages on S and T). The admissibility implies that $B = DE$ and $C = DF$, and the canonical forms constitute, in the terminology above, the direct product of the language of L with the free product of U and V. The formulas of Proposition 14.1 yield $A = D \cdot \frac{EF}{E+F-EF}$, which simplifies to the claimed formula.   QED

Note that the fact that the canonical forms are minimal can be interpreted as stating that both H and K are admissible in G. An admissible transversal for H, for example, is the set of alternating products $y = v_1 u_2 \cdots v_k$ (with $v_1 \neq 1$, unless $y = 1$, when the product is empty).

Proposition 14.3 ([Le 91, Ch 94b]) *Let (K, T) and (L, R) be isomorphic admissible subgroups of (H, S), with an isomorphism $\phi : K \to L$ mapping T on R. Let $G = \langle H, p \mid k^p = \phi(k) \ (k \in K) \rangle$ be the corresponding HNN-extension, with the generating set $\{S, p\}$, and let $A(X)$, $B(X)$, $C(X)$ be the generating growth functions of G, H, K, respectively. Then*

$$\frac{1}{A} = \frac{1}{B} - \frac{2X}{1+X} \cdot \frac{1}{C}.$$

Proof We choose admissible transversals U, V for K and L. Any element $x \in G$ can be written uniquely in the canonical form

$$x = hp^{n_1} a_1 p^{n_2} \cdots p^{n_m} a_m,$$

where $h \in H$, $n_i \neq 0$ is an integer, if $n_i > 0$ then $a_i \in V$, and if $n_i < 0$, then $a_i \in U$ ($m = 0$ is allowed) [LS 77, Theorem IV.2.1]. Consider a minimal form w of x, relative to the generators $\{S, p\}$. Collecting together elements of S, w becomes a product in which powers of p alternate with elements of H. We change it to a canonical form by collecting the elements of H to the left, as follows. We start with the rightmost occurrence, say h_n preceded by p^e. Assume that $e < 0$. If $h_n \in U$, we do nothing, and move left to the previous occurrence of an element from

H. Otherwise, write $h_n = ku$, with $k \in K$, $u \in U$, and then replace $p^e k$ by $\phi^{-e}(k)p^e$. Since ϕ maps T to R, it is length-preserving, and this move does not change the length of $p^e k$, and cannot increase the length of w. If $e > 0$, we write $h_n = lv$, and proceed similarly. Next we go one step left, to the previous occurrence of H, etc. This procedure leads to a canonical form, and shows that the canonical forms are minimal. We now count these forms.

First, $A(X) = B(X)D(X)$, where $D(X)$ is the generating growth function for canonical forms with $h = 1$. Note that in these forms a_m is the only factor that can take the value 1, and that the sign of n_i is determined by a_i, so we are allowed to choose only $|n_i|$, except when $a_m = 1$, when n_m can be either positive or negative. Therefore we count separately the words with $a_m = 1$ and the ones with $a_m \neq 1$, the resulting generating functions being denoted by D_1 and D_2.

If $a_m \neq 1$, then we have the free product of the language $\{p^e, e > 0\}$, and the language $U \cup V$, the latter without the empty word. The generating growth function of the first is $\frac{X}{1-X}$. For both U and V the corresponding function $E(X)$ equals $\frac{B(X)}{C(X)}$, and for $U \cup V$, the empty word removed, it is $2(E(X) - 1)$. We apply the argument of Proposition 14.1, i.e. of Proposition 1.5, which is simplified by the absence of the empty word in either language (in the notation of Proposition 1.5, the factors $B(X) - 1$ and $C(X) - 1$ are replaced by $B(X)$ and $C(X)$). This yields

$$D_2(X) = \frac{\frac{X}{1-X} \cdot (2E(X) - 2)}{1 - \frac{X}{1-X} \cdot (2E(X) - 2)} = \frac{X(2B - 2C)}{(1-X)C - X(2B - 2C)}$$
$$= \frac{X(2B - 2C)}{C + XC - 2BX}.$$

If $a_m = 1$, then we can regard the canonical forms as the direct product of the language $\{p^e, e \neq 0\}$ with the previous free product language, to which we add the empty word. The corresponding growth-generating functions are $\frac{2X}{1-X}$ and $D_2(X) + 1$, yielding, with the previous number of possibilities for the last factor p^{n_m} doubled, by a similar calculation

$$D_1(X) = \frac{2X}{1-X} \cdot \frac{C - XC}{C + XC - 2BX} = \frac{2XC}{C + XC - 2BX}.$$

Adding up, and adding 1 for the identity element, we have

$$A = B \cdot \frac{C - 2XB + XC + 2XC + X(2B - 2C)}{C + XC - 2BX} = B \cdot \frac{(1 + X)C}{C - 2XB + XC},$$

which is equivalent to the stated formula.   QED

As for amalgamated products, the minimality of the canonical forms implies that H is admissible in G.

As an illustration of the above results, we discuss *graph products*. Given a graph Γ, and groups A_γ, one for each vertex $\gamma \in \Gamma$, their *graph product* is the group $G = G(\Gamma) = \langle A_\gamma \mid [A_\gamma, A_\delta] = 1$, for each edge $\gamma\delta \in \Gamma \rangle$. The two extreme cases, where Γ is the complete graph, or the one with no edges, yield the direct and free products.

We now assume that Γ is finite, and the groups A_γ are finitely generated, by sets A_γ, say, and take $S := \bigcup S_\gamma$ as a generating set for G. Recall that a *full subgraph* of Γ consists of a subset of the vertices together with all the edges of Γ connecting them.

Proposition 14.4 *Let Δ be a full subgraph of Γ. Then $G(\Delta)$ is (isomorphic to) an admissible subgroup of $G(= G(\Gamma))$.*

Proof Let $H = \langle A_\delta \mid \delta \in \Delta \rangle$, a subgroup of G. Since the defining relations of $G(\Delta)$ are satisfied in H, there is an epimorphism $\phi : G(\Delta) \rightarrow H$. On the other hand there is an epimorphism $\psi : G \rightarrow G(\Delta)$, in which the subgroups $A_\gamma, \gamma \in \Delta$ are mapped onto themselves, and the other groups A_γ are mapped to 1. Since $\psi\phi = 1$, the map ϕ is 1-1, and $G(\Delta) \cong H$.

The proof of admissibility is by induction on the number of vertices of Γ. Obviously, we assume $\Delta \neq \Gamma$, and then there exists a vertex $v \notin \Delta$. Let E be the full subgraph on the vertices different from v. By induction, $G(\Delta)$ is admissible in $G(E)$. Let Z be the full subgraph on the vertices connected to v, and Y - the one on the same vertices and v. Then $G(Y) \cong G(Z) \times A_v$. Looking at the defining relations of G, we see that $G \cong G(E) *_{G(Z)} G(Y)$. Since $G(Z)$ is admissible in both $G(E)$ (by induction) and $G(Y)$, the remark following Proposition 14.2 shows that $G(E)$ is admissible in G, implying that $G(\Delta)$ is also admissible.   QED

Proposition 14.5 *([Le 91, Ch 94a]) In the notation above, write $A(X)$ and $B_\gamma(X)$ for the generating growth functions of G and A_γ respectively, and for each subgraph Δ, write $P_\Delta = \Pi_{\delta\in\Delta}(\frac{1}{A_\delta} - 1)$. Then*

$$\frac{1}{A(X)} = \sum P_\Delta(X).$$

The summation taken over all complete subgraphs (cliques) of Γ, including the empty one (for which $P_\emptyset = 1$).

Proof We apply the decomposition $G \cong G(E) *_{G(Z)} G(Y)$ obtained in the previous proof, where v is chosen as a vertex that is not connected to all other vertices. Then we can apply induction to E, Z, and H, and substitute the result in the formula of Proposition 14.2. The term corresponding to E accounts for all cliques not containing v, while the difference between the other two terms accounts for the cliques containing v.

There remains the possibility that all vertices are connected to all others, i.e. Γ is complete and G is the direct product of the groups A_γ. Then the summation extends over all subgraphs of Γ, and the equality is clear. QED

A particularly interesting case of graph products occurs when all the groups A_γ are infinite cyclic; then $G(\Gamma)$ is a finitely generated group whose defining relations are that certain pairs of the generators commute. These groups are known as *right-angled Artin groups*.

14.4 Central Products

Recall the definition. Let H, K be two groups with central subgroups, $X \leq H$ and $Y \leq K$, such that $X \cong Y$, and let $\phi : X \to Y$ be an isomorphism. Then $W = \{(x, \phi(x^{-1}))\}$ is a central subgroup of $X \times Y$, and $G = (H \times K)/W$ is the *central product* of H and K obtained by *identifying* X and Y. It contains $Z = (X \times Y)/W$ as a central subgroup isomorphic to X and Y, and there are natural isomorphisms from X and Y onto Z. We will deal only in the case that $Z = \langle z \rangle \cong \mathbb{Z}$ is infinite cyclic, and z is the image of generators x and y for X and Y. Given generating sets S, T for H, K, we take for G the generators $\{ST \cup S \cup T\}$. We write $L = H/X$, $M = K/Y$, $N = G/Z \cong L \times M$, and for an element $l \in L$ and a positive integer n, we fix a representative $r(l)$ of l in H, and write $C(l, n) = \{u \in \mathbb{Z} \mid l_H(r(l)x^u) \leq n\}$. We use similar notation for K and G, where in G we let $r(l)r(m)$ be the representative of $(l, m) \in N$. Then $s_H(n) = \sum_{l \in L} |C(l, n)|$ and $s_L(n) = |\{l \in L \mid C(l, n) \neq \emptyset\}|$, with similar formulas for K and G.

Proposition 14.6 *With the above notation, given $l \in L$ and $m \in M$, we have $C(lm, n) = C(l, n) + C(m, n)$.*

Proof First, let $u \in C(l, n)$ and $v \in C(m, n)$. That means $l_H(r(l)x^u) \leq n$ and $l_K(r(m)y^v) \leq n$. If we choose in $H \times K$ the same generators as in G, then

$$l_G(r(l)r(m)z^{u+v}) \leq l_{H \times K}(r(l)x^u, r(m)y^v) = \max(l_H(h), l_K(k)) \leq n$$

and $u + v \in C(lm, n)$.

Next, let $w \in C(lm, n)$. Then $l_G(r(l)r(m)z^w) \leq n$. We can find elements $h \in H$, $k \in K$ such that in G we have $r(l)r(m)z^w = hk$, and $l_{H \times K}(h, k) = \max(l_H(h), l_K(k)) = l_G(r(l)r(m)z^w)$. Looking at images in N shows that $h = r(l)x^u$ and $k = r(m)y^v$, for some $u, v \in \mathbb{Z}$, thus $w = u + v$, with $u \in C(l, n)$, $v \in C(m, n)$.   QED

Corollary 14.7

(a) $C(lm, n) \neq \emptyset$ iff $C(l, n) \neq \emptyset$ and $C(m, n) \neq \emptyset$.
(b) If $C(l, n)$ and $C(m, n)$ are intervals in $\mathbb{Z}$, of lengths r and s, then $C(lm, n)$ is an interval of length $r + s - 1$.

Theorem 14.8 *Continuing with the above notation, under the assumption of Proposition 14.7(b), we have $s_G(n) = s_H(n)s_M(n) + s_K(n)s_L(n) - s_L(n)s_M(n)$.*

Proof $s_G(n) = \sum |C(lm, n)| = \sum(|C(l, n)| + |C(m, n)| - 1)$, where the summation is over all pairs (l, m) such that $C(l, n) \neq \emptyset \neq C(m, n)$. For a fixed l, the summand $|C(l, n)|$ is counted once for each m such that $C(m, n) \neq \emptyset$, contributing $|C(l, n)|s_M(n)$, and adding over all l we obtain $s_H(n)s_M(n)$, and similarly we get the other two terms.   QED

Recall that with the above choice of generators we have $s_N(n) = s_L(n)s_M(n)$. Therefore the theorem can be written in the form

$$(14.1) \qquad \frac{s_G(n)}{s_N(n)} - 1 = \frac{s_H(n)}{s_L(n)} - 1 + \frac{s_K(n)}{s_M(n)} - 1.$$

Examples of groups in which $C(l, n)$ is an interval are provided in [St 96], where it is shown that the discrete Heisenberg group satisfies the assumptions of Proposition 14.7(b) (recall that the *discrete Heisenberg group* is the group $G = \langle x, y \mid [x, y, x] = [x, y, y] = 1 \rangle$, which can be described also as $U(3, \mathbb{Z})$, the group of upper unitriangular 3×3 matrices over $\mathbb{Z}$; the non-discrete Heisenberg group is the Lie group consisting of the same matrices over $\mathbb{R}$).

15
The Generating Function

We now want to discuss the nature of the generating growth function $A_G(X)$. We are interested in this function as an actual complex analytic function, not just a formal power series, and therefore we will write in this chapter the variable as z, not X. In the examples opening this book, $A(z)$ was usually a rational function. Assuming that this is the case, we write $A(z) = P(z)/Q(z)$, for some polynomials $P(z) = \sum p_n z^n$ and $Q(z) = \sum q_n z^n$, rewrite this as $A(z)Q(z) = P(z)$, and then rewrite this equation as the infinite set of equations $\sum a(k)q_{n-k} = p_n$ (where we take $p_n = 0$ for n bigger than the degree of P, and similarly for Q). We see that rationality of $A(z)$ is equivalent to the existence of linear recurrence relations for the numbers $a(n)$. We will refer to a group with a rational generating growth function as a *rational group*.

Theorem 15.1 *Let G be a rational group (relative to some set of generators). Then the coefficients of the growth-generating function can be taken as integers, and the growth of G is either polynomial or exponential.*

Proof Using the above notation, consider the recursion formulas for $a(n)$ as a system of linear equations for p_n and q_n. Since the $a(n)$s are integers, the system has a rational solution, and multiplying by a common denominator leads to an integral solution.

Next, decompose $Q(z)$, over $\mathbb{C}$, as a product of linear factors, $Q(z) = a\Pi(1 - \alpha_i z)$. Then $P(z)/Q(z)$ can be written as a sum of rational functions with denominators $(1 - \alpha_i z)^j$. Develop these denominators to infinite power series. The function $A(z)$ has poles at the points α_i^{-1}, and its radius of convergence ρ is the smallest absolute value of these numbers. If $\rho < 1$, then the growth is exponential, while if $\rho = 1$, then $|\alpha_i| \leq 1$

for all i, and this implies that the $a(n)$, which are linear combinations of powers α_i^j, grow polynomially. QED

Corollary 15.2 *There are only countably many rational groups.*

Nevertheless, there are many interesting rational groups. Sometimes this can be established by direct calculation – we saw some examples of that in the first and in the previous chapter, more often by general considerations. The formulas of the previous chapter show that in many natural constructions of new groups from old, if the old ones are rational, so are the new ones. This is clear for amalgamated products, HNN-extensions, and wreath products by a finite group. Example 2 of [Jo 91] shows that if G is rational, then so is $Gwr\mathbb{Z}$. For central products of the type discussed in Theorem 14.8, the rationality follows from that theorem and the following fact.

Proposition 15.3 *If $\sum a_n z^n$ and $\sum b_n z^n$ are power series with rational coefficients which represent rational functions, then $\sum a_n b_n z^n$ also represents a rational function.*

Proof Partial fractions decomposition shows that a power series $\sum c_n z^n$ represents a rational function iff there exist finitely many polynomials P_i and complex numbers β_i, such that $c_n = \sum_i P_i(n)\beta_i^n$, for n large enough. If a_n and b_n are of this form, then so is the product $a_n b_n$. Recall that we saw in the preceding proof that if a power series with rational coefficients represents a rational function over $\mathbb{C}$, it represents one over $\mathbb{Q}$. QED

Thus, e.g., if G is the discrete Heisenberg group and H is the central product of G by itself, then H has a rational growth series, with respect to the appropriate generators. As was mentioned already, it is shown in [St 96] that H has also a set of generators yielding a transcendental growth function.

Remark Given power series $A(z) = \sum a_n z^n$ and $B(z) = \sum b_n z^n$, the series $C(z) = \sum a_n b_n z^n$ is known as the *Hadamard product* of $A(z)$ and $B(z)$ (after J.S. Hadamard, 1865–1963, the prover of the *prime number theorem* (also proved by C.-J.È.G.N. de la Vallée Poussin, 1866–1962)).

Warning The analogue of 15.3 for algebraic functions need not hold! For example, $A(z) = \sum \binom{2n}{n} z^n = \frac{1}{\sqrt{1-4z}}$ is algebraic, but $C(z) = \sum \binom{2n}{n}^2 z^n$ is transcendental.

The generating growth functions of groups are power series with integer coefficients, and there are several results which restrict the behaviour of such series, when compared with general series. Recall that if an analytic function $f(z)$ is defined in a domain $\mathcal{D}$ bounded by a curve $\mathcal{C}$, then $\mathcal{C}$ is a *natural boundary* for $f(z)$, if $f(z)$ cannot be continued analytically to any domain containing $\mathcal{D}$ properly. The following holds:

Theorem 15.4 ([G. Pólya and F. Carlson [Re 98, p.265]]) *Let the analytic function $f(z)$ be represented by a power series with integral coefficients and radius of convergence 1. Then $f(z)$ is either rational, with denominator $(1 - z^m)^n$ (for some m, n), or transcendental, and in the latter case the unit circle is a natural boundary for $f(z)$.*

Combined with Theorem 15.1, this yields:

Corollary 15.5 *Let G be a group of subexponential growth. Then*

(a) *If G has intermediate growth, then its generating growth function is transcendental, and the unit circle is a natural boundary for it.*

(b) *If G has polynomial growth, then its generating growth function is either rational, with denominator $(1 - z^m)^n$ (for some m, n), or transcendental, and in the latter case it has the unit circle as a natural boundary.*

But, as mentioned above, there exist groups, necessarily of exponential growth, whose generating growth functions are irrational algebraic (see, e.g., [Pa 92], [FS 08]).

We indicate a possible application of the above results. Stoll's proof that some Heisenberg groups have transcendental generating growth function proceeds as follows:

Step 1 Prove that there is a number α and an integer d such that for the relevant group, say G, with a suitable set of generators, we have $a_n \sim \alpha n^d$.

This is a special case of a Theorem of Pansu that was mentioned in Chapter 4, but Stoll provides a simpler proof for the Heisenberg group.

Step 2 Prove that if $f(z)$ is a power series whose coefficients behave as in Step 1, and $f(z)$ is rational, or algebraic, then α is rational, or algebraic, respectively.

Step 3 Prove that α is transcendental.

This is done by evaluating α explicitly, obtaining $\alpha = A + B\log 2$, for some rational numbers A, B, and then quoting the fact that $\log 2$ is transcendental. But Corollary 15.5(b) shows that it suffices to apply the much easier fact that $\log 2$ is irrational.

Another relevant fact about power series is:

Theorem 15.6 ([Ti 39, 7.21]) *Let the analytic function $f(z)$ be represented by a power series with positive coefficients and radius of convergence R. Then $z = R$ is a singular point of $f(z)$.*

The generating growth function of $\mathbb{Z}$ is $f(z) = \frac{1+z}{1-z}$, and it satisfies the functional equations $f(-z) = \frac{1}{f(z)}$ and $f(\frac{1}{z}) = -f(z)$. It turns out that many other rational groups satisfy similar functional equations. Examples of that are the groups known as *Coxeter groups* or, more generally, *Fuchsian groups*. These can be defined either geometrically or algebraically, e.g. Coxeter groups are groups that are generated by finitely many elements, say $s_1, \ldots, s_d$, of order 2, the product of any two generators has finite order, and the equations $s_i^2 = 1$, $(s_i s_j)^{n_{i,j}} = 1$ are a presentation of the group. Besides the rationality and the functional equations, it often happens that the singular points of the generating growth functions have interesting arithmetical properties. Also, the values $f(1)$ (or $\frac{1}{f(1)}$) are related to other important invariants of the groups. For these results, see e.g. [FP 87], [So 06], or the introduction to [Sc 11]. The interested reader may follow other papers by the same authors, or papers quoted by, or quoting, these papers.

We discuss in detail finitely generated abelian groups. These are direct products of cyclic groups. The generating growth function of an infinite cyclic group is $\frac{1-z}{1+z}$, and for any finite group it is a polynomial. It follows from Example 8 on page 4 that all finitely generated abelian groups are rational. This, however, under the assumption that we consider a "natural" set of generators, i.e. take one generator for each infinite cyclic factor. But the result holds for all types of sets of generators. This can be proved directly, by combinatorial means, but we prefer a ring theoretical approach.

Recall that an *algebra* over a field F is a ring R, which is at the same time a vector space over F, such that the ring addition and the vector space addition in R are the same, and the multiplications are related by the identity $\alpha(xy) = (\alpha x)y = x(\alpha y)$, for $x, y \in R$ and $\alpha \in F$. The map $\alpha \to \alpha \cdot 1$ maps F isomorphically onto a subfield of R, and sometimes it is convenient to identify F with this subfield. This implies in particular that any module over R is automatically also a vector space over F.

In the same way one can define an algebra over any commutative ring, but we do not need this more general concept.

Definition A *graded algebra* over a field F is an algebra R over F, which can be decomposed as a direct sum $\bigoplus_0^\infty R_n$, where each R_n is a vector subspace of R, and $R_m R_n \le R_{m+n}$.

A *graded module* over R is a module M, which can be decomposed as $\bigoplus_0^\infty M_n$, such that $R_m M_n \le M_{m+n}$.

Note that R_0 is a subring of R, and that the other summands are modules over this subring. Also the submodules M_n are modules over R_0. These two observations can be unified by noting that R is a graded module over R_0.

One can define a more general type of graded rings and modules, where the indices of the direct summands can be taken from any commutative semigroup, but we do not need these.

An element of R that lies in some R_n, or an element of M that lies in some M_n, is termed a *homogeneous* element. Such an element, unless it is 0, is said to have *degree* n. A submodule is *homogeneous* if it is generated by homogeneous elements. Alternatively, a submodule N is homogeneous if it is a direct sum of subgroups of the M_n, i.e. $N = \bigoplus N \cap M_n$. This terminology is applied also to ideals of R, considered as submodules of R itself. If I is a homogeneous ideal, then the factor ring R/I is itself a graded ring, $R/I = \bigoplus R_n/(I \cap R_n)$, and if N is a homogeneous submodule, then both N and $M - N$ are graded modules over R.

Theorem 15.7 *Let R be a finitely generated commutative graded algebra, and let M be a graded module over R. Assume that each summand M_n is finite dimensional over F, of dimension d_n, say, and that $R_1 M_n = M_{n+1}$ for each n. Then the generating function $A(X) := \sum d_n z^n$ is a rational function.*

Proof Let R be generated by $a_1, \ldots, a_k$. We can write each a_i as a linear combination of homogeneous elements, and by replacing a_i by its components, we may assume that all generators are homogeneous. Let a_i have degree r_i. The proof is by induction on k. If $k = 1$, then $R_n \ne 0$ only if n is a multiple of r_1, and the assumption $R_1 M_n = M_{n+1}$ forces $a_1 \in R_1$ (or else $M = M_0$ and there is nothing to prove). Multiplication by a_1 is a homomorphism of M_n onto M_{n+1}, therefore the dimensions

d_n are a non-increasing sequence, hence they are constant from some point on, and rationality of $A(X)$ is clear for such a sequence. Now let $k \geq 2$, and let Φ_i be the endomorphism of M that is determined by multiplication by a_i. Then Φ_i maps M_n into M_{n+r_i}. Let K and L be the kernel and the image of Φ_k, respectively. Since a_k is homogeneous, so are K and L, say $K = \bigoplus K_n$ and $L = \bigoplus L_n$. Let e_n and f_n be the dimensions of K_n and of $M_n - L_n$. The ideal $a_k R$ is homogeneous, and since Φ_k is 0 on both K and M/L, both of these are graded modules over $R/a_k R$. The latter ring is generated by $k - 1$ elements, so by induction the series $\sum e_n z^n$ and $\sum f_n z^n$ are rational functions, say $B(z)$ and $C(z)$. Write $r = r_k$, then $L_{n+r} \cong M_n - K_n$, so $d_n = e_n + (d_{n+r} - f_{n+r})$, translated as $A(z) = B(z) + z^{-r}(A(z) - C(z) + P(z))$, where $P(z)$ is a polynomial composed of the first r terms of $A(z)$ and $C(z)$. Thus $A(z) = (z^r B(z) + (P(z) - C(z)))/(z^r - 1)$ is rational.   QED

Corollary 15.8 *Let R and M be as above. Then $A(z)$ can be written as a rational function with denominator $(1 - z^{r_1}) \cdots (1 - z^{r_k})$ and $d_n \leq C n^{k-1}$, for some constant C.*

Proof The shape of the denominator is clear from the above proof. Given that shape, $A(z)$ can be written as $P(z)Q_1(z) \cdots Q_k(z)$, for some polynomial P, and Q_i being the infinite geometric series for $1 - z^{r_i}$.

The coefficients in Q_i are either 0 or 1, and simple induction shows that the coefficients of the product are polynomials of degree at most $k - 1$.   QED

The following, while a trivial consequence of 15.7, is its main application, so we state it explicitly.

Corollary 15.9 *Let R be a finitely generated commutative graded algebra, such that the subspaces R_n are finite dimensional, of dimensions d_n, say, and such that R_1 and R_0 generate R (as a ring). Then the generating function $\sum d_n z^n$ is a rational function of z.*

Note that the assumption that R_1 and R_0 generate R is necessary to ensure that $R_1 R_n = R_{n+1}$ (actually the two assertions are equivalent).

The range of applicability of these results, as well as of other properties of graded rings, is enlarged by means of the following concept:

Definition An algebra R is *filtered* if $R = \bigcup_0^\infty R_n$, where each R_n is a vector subspace of R, and $R_n \leq R_{n+1}$ and $R_m R_n \leq R_{m+n}$.

Given a filtered algebra R, we associate to it a graded algebra S as follows. Let us write $S_n = R_n - R_{n-1}$, a vector space (R_{-1} is interpreted as 0), and $S = \bigoplus S_n$, also a vector space. We define multiplication in S. If $x \in S_m$ and $y \in S_n$, write $x = a + R_{m-1}$, $y = b + R_{n-1}$, where $a \in R_m$, $b \in R_n$. Define $xy = ab + R_{m+n-1}$, an element of S_{m+n}, and extend this definition to non-homogeneous elements by distributivity. A routine calculation verifies that S is an algebra, which is graded by the subspaces S_n, and is termed the *associated graded algebra* of R.

Exercise 15.1 Let R and S be a filtered ring and its associated graded ring, respectively. Show that if S has no zero divisors, then neither does R. Is the converse true?

Let G be any group (or even a semigroup) with a set of generators X, and let R be the group algebra FG. Thus FG is the set of formal linear combinations $\sum_{x \in G} \alpha_x x$, with $\alpha_x \in F$, and only finitely many coefficients are non-zero. Addition in FG is defined by adding the coefficients of the same group element x, and multiplication by the equation $(\alpha x)(\beta y) = (\alpha \beta)(xy)$, and distributivity. Writing R_n for the subspace of R spanned by the elements of G that can be written as words of length at most n in the given generators, R becomes a filtered algebra. Writing also V_n for the subspace spanned by the elements of length exactly n, we have $R_n = R_{n-1} \bigoplus V_n$. If X is finite, then the subspaces V_n are finite dimensional, and $\dim(R_n - R_{n-1}) = \dim V_n = a_G(n)$. Thus the last corollary yields the result alluded to earlier:

Theorem 15.10 *Let G be a finitely generated abelian group, fix any finite set of generators, consisting of d elements, say, and let a_n be the number of elements of length n relative to this generating set. Then the generating function $A(z) = \sum_0^\infty a_n z^n$ is a rational function, which can be written with denominator $(1 - z)^{2d}$.*

Note that the exponent is $2d$, because as ring generators for R we have to take both the elements of X and their inverses.

In [Be 83] and [Be 87] it is shown that the last theorem holds also for non-abelian groups that contain an abelian subgroup of finite index. But the theorem does not hold in nilpotent groups of class 2 [St 96]. Indeed, as we noted several times, for such groups it is possible that the generating growth function is rational relative to one set of generators, but transcendental relative to another set.

Another interesting class of rational groups is the class of *hyperbolic groups*. This notion was defined by Gromov and E. Rips (independently), as an algebraic analogue to metric spaces of negative curvature. Here we will deal only, as is our (fortunate or unfortunate) habit, with the purely algebraic aspects. As usual, we consider the Cayley graphs of a finitely generated group G as a metric space, and we look at *geodesic triangles* in this space, i.e. at three elements $x, y, z \in G$, and geodesic paths joining them in the Cayley graph. The basic model is free groups, whose Cayley graphs are trees, there is a unique geodesic between any two vertices, and all geodesic triangles are degenerate: one of the edges is just the union of the other two. We replace this degeneration by a weaker assumption: each point on each edge is close to some point on one of the other edges. Formally:

Definition A finitely generated group G is A-hyperbolic, for a positive number A, if in any geodesic triangle, any vertex on one of the edges is at distance at most A from some vertex on one of the other two edges. G is hyperbolic if it is A-hyperbolic for some A.

We also say that the triangles are *A-thin*, or just *thin*. Hyperbolic groups are also called *word hyperbolic*, *Gromov hyperbolic*, or *groups of negative curvature*. They have several equivalent definitions. For these, and for proofs of statements made below without either a proof or a reference, we refer, e.g., to [GH 90]. The definition above relates to a specific set of generators, but it turns out that hyperbolicity does not depend on the generating set. More generally, groups quasi-isometric to hyperbolic groups are themselves hyperbolic. In particular, the free group $\mathbb{Z}$ is hyperbolic, and the groups commensurable with it, i.e. the virtually cyclic groups, are termed *elementary hyperbolic groups*. Free abelian groups $\mathbb{Z}^k$, $k \geq 2$, whose Cayley graphs consist of all lattice points in k-dimensional Euclidean space, are not hyperbolic, while $\mathbb{Z}$ is hyperbolic, and thus hyperbolicity is not preserved in direct products, but it is preserved in free products. Since free groups are hyperbolic, hyperbolicity is not preserved in epimorphisms either. It is more difficult, but possible, to find examples of non-hyperbolic subgroups of hyperbolic groups [Ri 82].

Let us prove several simple, but basic, results.

Proposition 15.11 *Let G be an A-hyperbolic group, let $x \in G$, let $l(x) = n$, and let $x = u_1 \cdots u_n = v_1 \cdots v_n$ be two shortest representations of x as products of the generators and their inverses. Then for each $i \leq n$ we have $l(u_i v_i^{-1}) \leq 2A$.*

In a more geometric language, the proposition says that given any two geodesic paths in the Cayley graph of G from the identity to x, corresponding points on the two geodesics are at bounded distance from each other. Of course this holds also for any two geodesics with the same end points, since they are translates of geodesics starting at 1. Two such geodesics are said to be *fellow travellers*, and a finitely generated group G has the *fellow traveller property*, if there exists a number B (depending only on G and its given set of generators), such that any two geodesics with the same end points are fellow travellers, with distances bounded by B.

Proof of 15.11 Consider the triangle formed by 1, x, and v_i, with edges the path $1u_1 \cdots u_n$ and the segments $1v_1 \cdots v_i$ and $v_i \cdots v_n$. By hyperbolicity, the point u_i is at distance at most A from some v_j. If $j > i + A$, then $n - i = d(u_i, x) \leq d(u_i, v_j) + d(v_j, x) \leq A + (n - j) < n - i$: a contradiction. Thus $j \leq i + A$, and a symmetric argument shows that $i \leq j + A$, and $d(u_i, v_i) \leq d(u_i, v_j) + d(v_j, v_i) \leq A + |i - j| \leq 2A$.   QED

Theorem 15.12 *Groups with the fellow traveller property are finitely presented.*

Corollary 15.13 *Hyperbolic groups are finitely presented.*

Proof of 15.12 Let G have the fellow traveller property with bound B, and let it be generated by the finite set S. We will show that each equality $w = 1$ in G, where w is a reduced word on the letters in S, follows from similar equalities involving words of length at most $2B$. Thus the equalities of that length form a set of defining relations for G. Let $l(w) = n$ (the lengths of words are computed in the free group on S), assume that $n > 2B$, and apply induction on n. The equality $w = 1$ means that in the Cayley graph, following the word w letter by letter describes a closed path from the identity to itself. Let u be the middle letter of w (or one of the two middle letters). Thus u divides w into two parts, say w_1 and w_2. If w_1 is not a geodesic (from 1 to u), replace it by a geodesic, say v. Then $w = w_1 v^{-1} \cdot v w_2$, the two factors are closed paths shorter than w, and the induction applies. If w_1 and w_2 are geodesics, let u_1, u_2 be their middle points, dividing w_1 and w_2 into $w_3, \ldots, w_6$, and let v be a geodesic from u_1 to u_2. By assumption, $d(u_1, u_2) \leq B$, and this time we replace w by the closed paths $w_3 v w_5^{-1}$ and $w_4 w_6^{-1} v^{-1}$.   QED

Exercise 15.2 Using the assumptions and notation of of the above proof, show that w can be written (in the free group) as a product of at most 2^n conjugates of relations of length at most $2B$, with the conjugating elements having lengths at most n. Deduce that word problem is soluble in groups with the fellow traveller property.

Remark Often the bound 2^n can be replaced by a much smaller one. For hyperbolic groups, a linear bound obtains, and the existence of a linear bound characterizes hyperbolic groups. The conjugacy problem is also soluble in hyperbolic groups, and even the much more difficult isomorphism problem is soluble. This was noted first by Z. Sela for an important subclass [Se 95]. For the full result see [DG 11].

Hyperbolic groups are numerous among the finitely presented groups. More precisely:

Theorem 15.14 *For natural numbers $k, n_1, \ldots, n_r$, let $N(k, n_1, \ldots, n_r)$ be the number of groups that can be defined by k generators and r relations, with the relations of lengths $n_1, \ldots, n_r$, and let $N_h(k, n_1, \ldots, n_r)$ be the number of hyperbolic groups among them. Then*

$$\lim_{\min n_i \to \infty} \frac{N_h(k, n_1, \ldots, n_r)}{N(k, n_1, \ldots, n_r)} = 1.$$

This was observed by Gromov and proved by A.Olshanskii [Ol 92]. Three basic results about the growth of hyperbolic groups are:

Theorem 15.15 [Ko 98] *Non-elementary hyperbolic groups have uniformly exponential growth.*

This holds also for the larger class of *relatively hyperbolic* groups [Xi 07].

Theorem 15.16 *Hyperbolic groups have rational generating growth series with respect to any finite set of generators.*

Corollary 15.17 *The exponential growth rate of a hyperbolic group, with respect to any set of generators, is an algebraic integer.*

Indeed, Theorems 15.1 and 15.6, combined with the previous theorem, show that the growth rate is a root of a polynomial with integer coefficients, hence it is an algebraic number. Moreover, the proof in [GH 90] shows that that polynomial is the minimal polynomial of some integer matrix, hence the root is an algebraic integer.

16

The Growth of Free Products

In Chapter 14 we gave formulas for the growth functions of amalgamated free products and HNN-extensions. These, of course, can be explicitly given only in special situations. We now discuss the growth of all groups that can be so decomposed. The main result, due to Michelle Bucher and Pierre de la Harpe [HB 00] is that, with some explicitly described exceptions, these groups are not only of exponential, and even uniformly exponential, growth, but they form a family of groups with uniformly uniform exponential growth, i.e. there exists a uniform lower bound for the growth rate of all groups in the family. They give the explicit bound $\Omega(G) \geq \sqrt[4]{2} = 1.189\cdots$ for all groups in the family. As an application, Grigorchuk and de la Harpe [GH 01] showed that the same bound holds for the growth of one-relator groups, again with a few explicit exceptions, and J.O. Button [Bu 09] extended this to all groups with *positive deficiency*, i.e. ones that have a finite presentation with more generators than relators. Bucher [Bu 99] gave the sharper bound $\Omega(G) \geq \sqrt{2}$ for ordinary free products (excepting the infinite dihedral group $C_2 * C_2$, which has linear growth). Here, following [Ma 11], we will prove all these results, with improved bounds.

The relevant common aspect of amalgamated products and HNN-extensions is their actions on trees, and we start by describing these trees.

Let G be any group that can be generated by two proper subgroups, $G = \langle H, K \rangle$. Then we construct a graph $\Gamma := \Gamma(G, H, K)$, taking as vertices the right cosets of both H and K, and a coset Hx is connected by an edge to a coset Ky if $Hx \cap Ky \neq \emptyset$.

This is a so-called *bipartite graph*, i.e. the vertices are partitioned to two disjoint sets, and edges exist only between vertices from different sets. The group G acts on (the vertices of) Γ by right multiplication, and this action is transitive on each of the two components. A coset of H is connected to $|H : K \cap H|$ cosets of K, and a coset of K is connected to $|K : K \cap H|$ cosets of H. Thus the graph need not be locally finite. An extreme case occurs when $|H : H \cap K| = |K : H \cap K| = 2$. Then each vertex is connected to exactly two others, and this is possible if, and only if, Γ is an infinite straight line or a circle. In this case we say that the amalgamated product is of *dihedral type*. This is one of the exceptional cases that we have mentioned above. It includes, e.g., the infinite dihedral group, which has linear growth. For a discussion of the growth in this case, see [Br 05].

Proposition 16.1 *The graph Γ is connected. If G is an amalgamated product of H and K, $G = H *_L K$, then Γ is a tree.*

Proof To show connectedness, it suffices to show that the vertex H can be connected to any other vertex. Edges connect H to the cosets of K that intersect H, and these are the cosets that are contained in KH. These cosets, in turn, can be connected to the cosets of H that are contained in HKH, and, continuing, we see that H can be connected in n steps to the cosets that are contained in an alternating product of H and K of the form either $HKHK \cdots H$, or $KHK \cdots H$, whichever has n factors. Each coset of H and K is contained in such a product, therefore the graph is connected.

Now let G be an amalgamated product. To prove that Γ is a tree, we have yet to show that it contains no cycles. We apply the notations that we used in the proof of Proposition 14.2. Thus each element of G is uniquely expressible as a product $x = lu_1 v_1 \cdots u_n v_n$, with $l \in L$, and u_i, v_i are taken from sets of representatives of the right cosets of L in H and K. If this x is a representative of the coset Hx, then, since $lu_1 \in H$, the element $v_1 \cdots u_n v_n$ represents the same coset. Thus each coset of H has a representative of the form $x = v_1 \cdots u_n v_n$, and if $x \neq 1$, then $v_1 \neq 1$; similarly for K. Consider a path in Γ. We may assume that it starts at H. The next vertex in the path is a coset of K with representative u_n, and $u_n = 1$ is possible. In the next step we have a coset of H intersecting Ku_n, so we can find a representative of the form ku_n, with $k \in K$; and writing $k = lv_n$, with $l \in L$, we can take $v_n u_n$ as the representative, and we must have $v_n \neq 1$, otherwise we go back to H. In the same way, from now on, going along the cycle, at each stage we multiply by either some u_i or v_i, which are not 1, and thus we never get as representative an element of H, i.e. the path never closes, Γ contains no cycles, and it is a tree. QED

We note in passing that this argument implies also the converse: *if Γ is a tree, then G is an amalgamated product of H and K.*

Next, let $G = \langle H, p \rangle$, with H a subgroup, and p an element, of G. We define a directed graph $\Delta = \Delta(G, H, p)$. The vertices are the cosets Hx, and there is a directed edge leading from Hx to Hy if $Hx \cap pHy \neq \emptyset$. This time the group G is transitive on Δ. Let $K\ (= H \cap H^{p^{-1}})$ be the largest subgroup of H such that $K^p \leq H$, and let $L = K^p$. There is an edge from H, say, to Hy if Hy has a representative in $p^{-1}H$, and if two elements from that left coset, say $p^{-1}h_1$ and $p^{-1}h_2$, define the same right coset, that means that for some $h \in H$ we have $p^{-1}h_1 = hp^{-1}h_2$, i.e. $p^{-1}h_1 h_2^{-1} p \in H$, and $h_1 h_2^{-1} \in K$. Therefore the numbers of edges leading out from each vertex of Δ is $|H : K|$, and similarly, the number of edges entering each vertex is $|H : L|$. It follows that Δ is a straight line or a circle only if $K = L = H$, i.e. G is a semidirect product of H by $\langle p \rangle$, and this is the exceptional case now.

Proposition 16.2 *The directed graph Δ is connected, and if G is an HNN-extension of H, with stable letter p, then Δ is a tree.*

Proof There is an edge from Hx to Hy iff $y \in Hp^{-1}Hx$, and from Hy to Hx iff $y \in HpHx$. Writing y as a product of elements of h and powers of p, we see that there is some x, satisfying one of the two inclusions, that can be written as a similar product with fewer factors. Induction on this number of factors shows that each vertex can be connected to H, and the graph is connected.

Now let G be an HNN-extension. In the proof of Proposition 14.3 we noted that each element of G can be written uniquely in the form $x = hp^{n_1} a_1 p^{n_2} \cdots p^{n_m} a_m$, where $h, a_i \in H$. We write this in the form $x = hp^{\pm 1} z$. For each coset of H we can find a representative in that form with $h = 1$, and from the previous paragraph, if Hy is connected to Hx, then y can be written as $p^\epsilon bx$, with $\epsilon = \pm 1$ and $b \in H$. If Hx and Hy are adjacent vertices in a path without backtracking, we may assume that the other coset adjacent to Hx in that path is Hz. Then either $b \neq 1$ or the signs of ϵ and n_1 are opposite, otherwise there is a backtracking. Thus we see that in such a path, starting from H, say, the lengths of the representatives grow, and we never return to H, so there are no cycles.   QED

Having defined the trees and the action on them, we need to discuss automorphisms of trees. Such an automorphism is an *inversion*, if it interchanges the two vertices of some edge. It is clear that in both cases that we discussed, the group G contains no inversion. This is because in Γ two connected vertices belong to distinct orbits of G, and Δ is directed, with the action of G preserving the direction.

Definition An automorphism of a tree is *hyperbolic*, if it fixes no vertex or edge.

Proposition 16.3 *Let σ be a hyperbolic automorphism of a tree T, let n be the minimal distance between any vertex of T and its σ image, and let L be the set of vertices x such that $d(x, \sigma(x)) = n$. Then L is an infinite straight line.*

Proof Let $u \in L$. Clearly, also $\sigma(u) \in L$. Moreover, the whole geodesic connecting u to $\sigma(u)$ is transformed into the one connecting $\sigma(u)$ with $\sigma^2(u)$, and it follows that for each point v in that geodesic we have $d(v, \sigma(v)) = d(u, \sigma(u))$, and thus $v \in L$. Continuing like this, we see that the full orbit of that geodesic under all powers of σ, including the negative powers, is a straight line, say M, contained in L. Let $w \notin M$, and let u be the point of M nearest to w. Then the path leading from w to $\sigma(w)$ via the shortest route to u, then through M to $\sigma(u)$, and finally from $\sigma(u)$ to $\sigma(w)$ by the shortest route, has no backtracking, and since T is a tree, this path is the geodesic between w and $\sigma(w)$, showing that $d(w, \sigma(w)) > n$. Thus $L = M$. QED

Definition Let σ be a hyperbolic automorphism of a tree. The line L that was just shown to exist is called the *axis* of σ.

Let us label the vertices of L using integers. We can do it in such a way that if $u_i \in L$, then $\sigma(u_i) = u_{i+n}$. With this labelling, let the *right half tree* R_σ be the set of vertices w such that the nearest point to them on L is u_k, with $k > 0$ (the terminology comes from pretending that L is the x-axis in the plane). Note that this half tree is σ-invariant.

Theorem 16.4 ([HB 00]) *Let σ and τ be two hyperbolic transformations of the tree T, with different axes. Then among the four automorphisms σ, τ, σ^{-1}, τ^{-1}, two generate a free monoid.*

Proof Let L and M be the two axes. We label them (with integers) in a special way. If they are disjoint, we choose $u \in L$ and $v \in M$ such that $d(u, v)$ is the minimal distance between points of L and points of M; we label these two points as u_0 and v_0, and label the other points so that, as above, $\sigma(u_i) = u_{i+n}$ and $\tau(v_i) = v_{i+m}$, for some positive numbers n, m. If L and M intersect, then, choosing any labelling of L, there is a point

$u_i \in L \cap M$ such that $u_{i+1} \notin M$. We change the labelling, renaming u_i as u_0, and also as v_0, and renaming u_{i+1} as u_1. This fixes the labelling on L, and because $\mathcal{T}$ is a tree, no point u_r, $r > 0$, lies on M. We can also label the points of M in such a way that no v_s, $s > 0$, lies on L. Replacing either of σ, τ, or both, by their inverses, if necessary, we may again assume that $\sigma(u_i) = u_{i+n}$ and $\tau(v_i) = v_{i+m}$.

Let R_σ and R_τ be the corresponding half trees. We first claim that they are disjoint. Indeed, if $w \in R_\sigma \cap R_\tau$, let u_i and v_j be the nearest vertices to w on L and M, take the geodesics connecting w to them, extend them by the segments of the axes leading to u_0 and v_0, and then connect u_0 to v_0, if necessary. We obtain a closed circuit in $\mathcal{T}$, which is impossible. Moreover, let $w \in R_\tau$. Then connecting it to v_j as before, then along M to v_0 and u_0, defines a path without backtracking in $\mathcal{T}$; therefore it is a geodesic, showing that u_0 is the nearest point to w on L. Then the nearest point to $\sigma(w)$ is $\sigma(u_0) = u_n$, i.e. $\sigma(w) \in R_\sigma$. We thus have $\sigma(R_\sigma \cup R_\tau) \subseteq R_\sigma$, and similarly $\tau(R_\sigma \cup R_\tau) \subseteq R_\tau$.

We now carry out an argument that is known as "playing ping-pong". To show that σ and τ generate a free monoid, let $w(\sigma, \tau)$ be any word, formed only with positive powers of the generators. Suppose that it starts with σ^k, $k > 0$. Then $w(R_\sigma \cup R_\tau) \subseteq R_\sigma$. In particular, $w(R_\tau) \subseteq R_\sigma$, and since $R_\sigma \cap R_\tau = \emptyset$, we cannot have $w = 1$. Similarly for words starting with a power of τ.   QED

A free monoid of rank 2 has 2^n elements of length n. To show that a group G that is either an amalgamated free product or an HNN-extension has uniformly exponential growth, it will suffice to find two elements of G that in the action of G on a tree induce hyperbolic transformations with different axes, and to bound the lengths of these two elements. For the first task, it usually suffices to find just one hyperbolic element, say x. That is because, if the tree is not a straight line, the transitivity properties of $\mathcal{T}$ show that some generator, say a, does not preserve L_x, and then x and x^a induce hyperbolic automorphisms with the distinct axes L_x and $a(L_x)$. Moreover, $l(x^a) \le l(x) + 2$, so that it suffices to bound $l(x)$. We see that we have to exempt the cases when $\mathcal{T}$ is a straight line, i.e. amalgamated products of dihedral type and semidirect products.

Bounding the length of the hyperbolic elements is enabled by the following result:

Theorem 16.5 ([Se 80]) *If a group G, finitely generated by S, acts on*

a tree, and each element, or a product of two elements, of S, has a fixed point, then G fixes a point.

First:

Lemma 16.6 *Let σ act on a tree T with fixed points, and let u be a non-fixed vertex. Then $d(u, \sigma(u))$ is even, and the middle point of a geodesic between u and $\sigma(u)$ is fixed by σ.*

Proof Let v be the nearest fixed point to u, and let w be the next point to it on a geodesic L from u to v. Then $\sigma(w) \neq w$. Therefore on the path $L \cup \sigma(L)$ there is no backtracking, and thus this path is the geodesic from u to $\sigma(u)$, and its midpoint v is fixed by σ.   QED

Proof of 16.5 By induction on $|S|$, the case of one generator being trivial. Separate S to two disjoint proper subsets, $S = A \cup B$. By induction, the set X of all vertices fixed by A, and the set Y of all vertices fixed by B, are not empty. We need to show that $X \cap Y \neq \emptyset$. Suppose that X and Y are disjoint, let $u \in X$, $v \in Y$ yield the minimal distance between points of X and Y, and let w be the vertex next to v on the geodesic L from u to v. Let $\tau \in B$. Then $\tau(w) \neq w$, and as in the lemma, $L \cup \tau(L)$ is a geodesic from u to $\tau(u)$. But for each $\sigma \in A$ we have $(\tau\sigma)(u) = \tau(u)$, and therefore $L \cup \tau(L)$ is also the geodesic from u to $(\tau\sigma)(u)$, and v, the midpoint of that geodesic, is fixed by $\tau\sigma$, and also by σ: a contradiction.   QED

If G is either an amalgamated free product, not of dihedral type, or an HNN-extension, not a semidirect product, acting on the graphs Γ or Δ described above, then clearly G has no fixed points, and Theorem 16.5 shows that given any set of generators, there exists an element x of length at most 2 acting hyperbolically. Then for some generator a, x and x^a have length at most 4, and they (or their inverses) generate a free semigroup, which contains 2^n elements of length n in x and x^a, and therefore of length at most $4n$ in the generators of G. Thus $s_G(4n) \geq 2^n$, implying $\Omega(G) \geq \sqrt[4]{2} = 1.189\cdots$. Thus the family of amalgamated free products and HNN-extensions, exempting the products of dihedral type and the semidirect products, are of uniformly uniform exponential growth.

We want to improve the lower bound for $\Omega(G)$. For that, we have to separate the two cases. Let $\zeta = \frac{1+\sqrt{5}}{2} = 1.618\cdots$ be the famous *golden ratio*.

Theorem 16.7 *Let G be an HNN-extension, not a semidirect product. Then $\Omega(G) \geq \zeta$.*

Before giving the proof, we indicate two other ways to improve the bound $\sqrt[4]{2}$. These yield weaker bounds, but are easier to derive.

First, the stabilizers of vertices of the tree $\Delta(G, H, p)$ are the subgroup H and its conjugate. They are contained in the normal closure H^G, which is the kernel of the natural homomorphism of G onto $\langle p \rangle \cong \mathbb{Z}$. Since H^G is a proper subgroup, in any generating set S there is at least one element x which lies outside H^G, and thus does not fix any vertex: in other words, it induces a hyperbolic transformation of Δ. We thus can find, as above, two hyperbolic elements, say s and s' (x and x^a, or their inverses) of lengths 1 and 3 (at most). Words of length n in s and s' have length $3n$ at most in G, leading to the inequalities $s_G(3n) \geq 2^n$ and $\Omega(G) \geq \sqrt[3]{2} = 1.260 \cdots$.   QED

A further improvement is obtained by taking into account that the two hyperbolic elements have different lengths. Let M be the free monoid that s and s' generate.

Lemma 16.8 *Let $y \in M$ have length n in M. Then $l_G(y) \leq 2n + 1$. Equality is possible only if y, written as a reduced word in s, s', ends in s'.*

Proof By induction on n. The case $n = 1$ is clear. Let $x = uv$ be an element of length n in M, where u has M-length $n-1$ and v is s or s', and has S-length 1 or 3. By induction, $l_G(u) \leq 2n-1$. If $v = s$, it follows that $l_G(x) \leq 2n$. Assume that $v = s'$. If $l_G(u) \leq 2n - 2$, we are done. Let $l_G(u) = 2n - 1$. Then, again by induction, u ends in s'. Written in terms of S, that means that u ends in a, while v starts with a^{-1}, and thus there is a cancellation in uv, and $l_G(x) \leq 2n$.   QED

The lemma implies that $s_G(2n+1) \geq 2^n$, and $\Omega(G) \geq \sqrt{2} = 1.414 \cdots$.

Proof of Theorem 16.7. We estimate s_n more carefully. Let, as above, $x \notin H^G$ be a hyperbolic generator, with axis L, and let y be a generator which does not fix L. If y is also hyperbolic, its axis is not L, and then x and y, or x and y^{-1}, generate a free monoid, and this shows that the rate of growth is at least 2. If y is not hyperbolic, it belongs to some conjugate of H, and xy is hyperbolic and does not fix L. Assume first that x and xy generate a free monoid. Let F_n be the number of elements of G-length n in that monoid. An element of length n is obtained by writing either x or xy to the right of a shorter word, and this shows that $F_n = F_{n-1} + F_{n-2}$. Since the monoid has one element of length 1 and two of length 2, F_n is the nth Fibonacci number, and $\omega(G, S) \geq \lim F_n^{1/n} = \lim F_{n+1}/F_n = \zeta$. The other possibility is that x^{-1} and xy generate a free monoid. That monoid contains $x^{-1}xy = y$, each positive word in x^{-1} and y can be written as a positive word in x^{-1} and xy, and therefore x^{-1} and y also generate a free submonoid, and the growth rate is at least 2.   QED

We indicate a slightly different approach, both for its independent interest, and because it is needed later.

Proposition 16.9 *Let F be a free group freely generated by s and t, let K be the free submonoid generated by s and s^t, and let d_n be the number of elements of K length n in F. Then $\limsup d_n^{1/n} = \zeta$.*

Proof In this proof we write $l(x)$ for the length in F of the element x of K. Such an element is a product of terms equal to s or to s^t. Write b_n, c_n for the number of elements of K of length n which, as words in K, end in s or s^t, respectively, and let $d_n = b_n + c_n$ be the total number of elements of K of length n. An element x contributing to b_n is obtained simply by appending s to the right of a shorter word, and therefore $b_n = d_{n-1}$. An element contributing to c_n has the form $x = yt^{-1}st$. Here, if y ends in s, then $l(y) = n - 3$, and the number of possibilities for these ys is $b_{n-3} = d_{n-4}$. On the other hand, if y ends in $t^{-1}st$, then in $x = ys^t$ there is a cancellation of the pair tt^{-1}. That means that $l(y) = n - 1$ and the number of possible ys is c_{n-1}. Thus

$$c_n = d_{n-4} + c_{n-1} = d_{n-4} + d_{n-1} - b_{n-1} = d_{n-4} + d_{n-1} - d_{n-2}$$

and

$$d_n = b_n + c_n = 2d_{n-1} - d_{n-2} + d_{n-4}.$$

Let $D(z) = \sum d_n z^n$. Then $D(z) = P(z) + 2zD(z) - z^2 D(z) + z^4 D(z)$, where the polynomial $P(z)$ compensates for the fact that the recursion formula for d_n is valid only for $n \geq 4$. Thus

$$D(z) = \frac{-P(z)}{z^4 - z^2 + 2z - 1}.$$

The radius of convergence of $D(z)$ is the least absolute value of a singularity of $D(z)$, i.e. the least absolute value of a root of the denominator, and since the coefficients of $D(z)$ are positive, that singularity and root are also positive. Now $\limsup d_n^{1/n}$ is the reciprocal of that root, i.e. a root of the "reciprocal" polynomial

$$z^4 - 2z^3 + z^2 - 1 = z^2(z-1)^2 - 1 = (z(z-1)+1)(z(z-1)-1)$$
$$= (z^2 - z + 1)(z^2 - z - 1).$$

Thus the root in question is the golden ratio.

There is a mapping $\phi :\ K \rightarrow M$, mapping s and s^t to s and s', respectively. Since M is free, this mapping is 1-1, and $l_S(\phi(x)) \leq l(x)$. Thus $s_n \geq d_n$, and

$$\omega(G, S) = \lim s_n^{1/n} \geq \limsup d_n^{1/n} = \zeta,$$

proving Theorem 16.7.   QED

The sequence $\{d_n\}$ is sequence number A005252 in [Sl], with the first two terms omitted. Since $c_n = d_n - d_{n-1}$, the sequence $\{c_n\}$ is the difference sequence of $\{d_n\}$, and satisfies the same recursion as $\{d_n\}$, but has initial values 0,0,1,2, while $\{d_n\}$ starts with 1,1,2,4. The same recursion, with initial values 1,1,2,3, defines the Fibonacci sequence. Thus the generating functions of our sequences and of the Fibonacci sequence are rational functions with the same denominator, which explains why they have the same rate of exponential growth. The sequence $\{c_n\}$, with the two initial zeroes omitted, is sequence number A024490 in [Sl].

Theorem 16.10 *Let G be a non-trivial amalgamated free product, not of dihedral type. Then $\Omega(G) \geq \alpha$, where $\alpha = 1.324 \cdots$ is the unique positive solution to $z^3 - z - 1 = 0$.*

Proof This time we cannot guarantee hyperbolic elements of length 1, but we can find elements s, s' as before, such that $l_G(s) \leq 2$ and $l_G(s') \leq 4$, and they generate a free submonoid M of G. There is a lemma analogous to 16.8.

Lemma 16.11 *Let $x \in M$ have length n in M. Then $l_S(x) \leq 3n + 1$. Equality is possible only if x, written as a reduced word in s, s', ends in s'.*

The proof is identical to the one of Lemma 16.8. The lemma implies that $s_{3n+1} \geq 2^n$, and $\omega(G, S) \geq \sqrt[3]{2} = 1.260 \cdots$. To prove Theorem 16.10 in full, we estimate s_n as before. Let F be a free group, freely generated by u, v, w. Write $s = uv$ and $s' = s^w$. Then s and s' generate a free submonoid K of F. There is an isomorphism $\phi :\ K \rightarrow M$, sending s and s' onto the elements with the same name in G, and $l_S(\phi(x)) \leq l_F(x)$. We want to find the number of elements of F-length n in K.

Note that s and s' have lengths 2 and 4, and each word in them has an even length, so we consider only even lengths. We write b_n, c_n for the numbers of elements of K of F-length n which end in s or s', respectively, and $d_n = b_n + c_n$ for the total number of elements of K of length n. An element contributing to b_n is obtained by appending

$s = uv$ to the right of a shorter word, and therefore $b_n = d_{n-2}$. For c_n we obtain, as in the proof of Proposition 16.9, $c_n = b_{n-4} + c_{n-2} = d_{n-6} + c_{n-2}$, and $d_n = d_{n-2} + d_{n-6} + c_{n-2} = 2d_{n-2} - d_{n-4} + d_{n-6}$. Let $F(z) = \sum d_n z^n$ be the generating function of $\{d_n\}$. The recursion implies that $F(z) = P(z) + 2z^2 F(z) - z^4 F(z) + z^6 F(z)$, for some polynomial $P(z)$. Thus $F(z) = \frac{P(z)}{1 - 2z^2 + z^4 - z^6}$. As before, $\limsup d_n^{1/n}$ is the unique positive root of the polynomial reciprocal to the denominator, i.e. $z^6 - 2z^4 + z^2 - 1 = z^2(z^2 - 1)^2 - 1 = (z^3 - z + 1)(z^3 - z - 1)$. This proves Theorem 16.10.[1]   QED

Sharper inequalities can be given for ordinary free products.

Theorem 16.12 *Let $G = A * B$ be a non-trivial free product, other than the infinite dihedral group $C_2 * C_2$.*

(a) $\Omega(G) \geq \sqrt{2}$, *and unless $G = C_2 * C_3$, we have $\Omega(G) \geq \beta$, where $\beta = 1.544\cdots$ is the positive root of $z^3 - 2z^2 + 2z - 2 = 0$.*

(b) *If neither A nor B can be generated by elements of order 2, or if A cannot be generated by elements of order 2 and 3, then $\Omega(G) \geq \zeta$.*

(c) *Unless $G = C_2 * C_3$, or both A and B are C_2 or dihedral, $\Omega(G) \geq \gamma$, where $\gamma = 1.574\cdots$ is the positive root of $z^3 - z^2 + z - 3 = 0$.*

Before the proof, we make several comments. First, if $G = C_2 * C_3$ ($\cong \mathrm{PSL}(2, \mathbb{Z})$), and we take as the set S of generators the two generators of the cyclic free factors, then $\omega(G, S) = \sqrt{2}$ (see Example 10 of Chapter 1 on page 7). Thus $\Omega(C_2 * C_3) = \sqrt{2}$. In Section 14.1 we saw that the trefoil group $H = \langle x, y \mid x^2 = y^3 \rangle$ has a set of generators S, relative to which $\omega(H, S) = \sqrt{2}$ holds. There is a homomorphism from H onto G, therefore $\Omega(H) \geq \Omega(G)$, and thus $\Omega(H) = \sqrt{2}$. Moreover, let $H_k = \langle x, y \mid x^2 = y^3, x^{2k} = 1 \rangle$, then there are homomorphisms from H onto H_k, and from H_k onto G, hence $\Omega(H_k) = \sqrt{2}$ as well. Note that H_k is an amalgamated free product $C_{2k} *_C C_{3k}$, where $C \cong C_k$.

Next, Theorem 16.12 applies to many amalgamated free products, via the natural epimorphism $A *_C B \to A/C^A * B/C^B$.

Finally, using the natural generators for $G = C_2 * C_4$, it is easy to see that $\omega(G, S) = \zeta$, and Theorem 16.12(c) implies that $\Omega(G) = \zeta$. Let $H = \langle x, y \mid x^2 = y^4 \rangle$. We remarked in Section 14.1 that $\omega(H, \{x, y\}) = \zeta$. Since there is an epimorphism from H onto G, we obtain $\Omega(H) = \zeta$, and it follows that $\Omega(J_k) = \zeta$, where $J_k = \langle x, y \mid x^2 = y^4, x^{2k} = 1 \rangle = C_{2k} *_C C_{4k}$. Note also that if $G = C_2 * (C_2 \times C_2)$, with the natural

---

[1] $\{d_{2n}\}$ is sequence no. A005251 in [Sl], immediately preceding the one of Theorem 16.7.

set S of three generators, then $\omega(G, S) = \zeta$, but it is not known if $\Omega(G) = \zeta$. There are also no known examples of equality in Theorems 16.7 and 16.10.

Proof of Theorem 16.12   Let $G = A * B$ be generated by S. We let G act on the tree $T = \Gamma(G, A, B)$. There are two cases.

1. *Some element $s \in S$ is hyperbolic.* Then there exists another generator $t \in S$ such that s and $s' := t^{-1}s^e t$ ($e = \pm 1$) generate a free submonoid M of G. Lemma 16.8 applies, and implies that $s_{2n+1} \geq 2^n$, and $\omega(G, S) \geq \sqrt{2}$, the desired conclusion, but actually Proposition 16.9 shows that $\omega(G, S) \geq \zeta$ in this case.

2. *All elements of S fix some vertex.* The stabilizer of Ax is A^x, and the stabilizer of By is B^y. Thus our assumption means that each generator belongs to a conjugate of either A or B. Since the normal closures of A and B are proper subgroups, there must be generators lying in conjugates of both A and B. Write $S = T \cup U$, where T and U consist of the generators lying in conjugates of A and B, respectively. Choose generators $t \in T$ and $u \in U$. In the natural projection $\pi \colon G \to B$, tu maps onto a non-identity element of B, therefore $tu \notin Ker(\pi) = A^G$, and certainly tu does not lie in any conjugate of A. Similarly, tu does not lie in any conjugate of B, i.e. tu does not fix any vertex of the tree X, and it induces on it a hyperbolic automorphism. Let L_{tu} be the axis of that automorphism. If we can find a similar pair $\{v, w\}$ of generators, such that $L_{vw} \neq L_{tu}$, then tu and vw (or their inverses) generate a free submonoid, and this implies that $s_{2n} \geq 2^n$, and $\omega(G, S) \geq \sqrt{2}$. Thus we may assume that all products like tu have the same axis: in particular, $L_{tu} = L_{ut}$. But t and u^{-1} transform tu to ut, therefore they preserve L_{tu}. Each generator can play the role of either u or t, therefore all generators preserve the common axis L_{tu}. Recall that X is bipartite, that G has two orbits on it, and that vertices at distance 1 belong to different orbits. Therefore both orbits are represented in L_{tu}, and since that line is fixed by all generators, and therefore by G, it follows that $X = L_{tu}$ is a straight line, which is possible only if G is the infinite dihedral group $C_2 * C_2$.

That proves the inequality $\Omega(G) \geq \sqrt{2}$. For the rest of the proof, we may assume that we are in case **2**, and distinguish several subcases. We choose t and u as above.

2a. *Neither T nor U consists only of elements of order 2.* We may assume that t and u are different from their inverses, and then form 2^n products of length n, in which t and t^{-1} alternate with u and u^{-1}. Thus $\omega(G, S) \geq 2$ in this case.

From now on we assume that T consists of involutions.

2b. *u can be chosen to have order neither 2 nor 3.* Then u and t generate a free product $H := C * C_2$, where C is cyclic of order at least 4, and it is clear that the growth function of H, relative to $\{u, t\}$, is not less than the growth function of $C_4 * C_2$, relative to the natural generators, say u and v. Thus $\omega(G, S) \geq \omega(H, \{u, t\}) \geq \omega(C_4 * C_2, \{u, v\})$, and it was already noted that the last number is ζ.

We have by now proved Theorem 16.12(b). We now assume that T consists of involutions, and U consists of elements of order 2 or 3. It follows that A is generated by involutions, and B is generated by elements of order 2 or 3. We need the following:[2]

Lemma 16.13 *Let $G = A * B$ be a free product, and let s_α be a set of elements, which project onto distinct elements in the natural map $G \to A$. Then the subgroups B^{s_α} generate their free product in G.*

Proof Write an element of $\langle B^{s_\alpha} \rangle$ as $x = \prod b_i^{s_i}$, with $b_i \neq 1$ and $s_i \neq s_{i+1}$. We have to show that $x \neq 1$. We may assume that the identity element is one of the s_α. Write $l(\alpha) = l(s_\alpha)$ (in this proof, we use $l(x)$ for the free product length). If $x = 1$, then some b_i must be cancelled in the product (otherwise, when performing cancellations in x, between any two occurrences of b_i there stands a word which projects non-trivially on A, and complete cancellation is impossible). The only possibility for this to happen is that $s_i = 1$, s_{i-1} ends with c, and s_{i+1} ends with d, with $c, d \in B$, and $c b_i d^{-1} = 1$ (one of c, d may be 1, but not both). Thus $s_{i-1} = yc$, $s_{i+1} = zd$, and we can replace in x the segment $b_{i-1}^{s_{i-1}} b_i^{s_i} b_{i+1}^{s_{i+1}}$ by $c^{-1} \cdot y^{-1} b_{i-1} y \cdot z^{-1} b_{i+1} z \cdot d$, for some $y, z \in G$, where the projections of y, z on A are equal to the projections of s_i, s_{i-1}, and $l(y) + l(z) < l(s_{i-1}) + l(s_{i+1})$. We can employ induction on $\sum l(s_i)$ to finish the proof.   QED

2c. *We may choose u of order 3, and A is not cyclic (i.e. it is not of order 2).* We may assume that $u \in B$. The generators in T map, in the natural projection $G \to A$, onto generators of A; therefore there are at least two of them, say s and t, which map onto different non-identity elements of A. By the lemma, B^G contains the free product $B * B^s * B^t$. We consider products $w := u^{\pm 1}(u^{\pm 1})^y$, where y is s or t. There are 8

---

[2] In the corresponding point in [Ma 11] (the paragraph between 2.b and 2.c in the proof of Theorem 4) the Kurosh subgroup theorem is misquoted. Lemma 16.13 corrects that gap in the proof.

such words, of length 4. Forming products of n such words, we obtain 8^n words of length $4n$, and thus $w(G, S) \geq (8^n)^{1/4n} = 2^{3/4} > \zeta$.

Therefore if $\Omega(G) < \zeta$, then either $A \cong C_2$, or both A and B can be generated by involutions.

2d. B *is neither* C_3 *nor dihedral.* Then S contains elements t, u, v, w, where $t \in T$ (we may assume that $t \in A$), and u, v, w are elements of U mapping onto distinct non-identity elements in the natural projection $G \to B$. Here we take u, v, w to be either three involutions, or u and v have order 3 and are inverse to each other. Then G contains the free product of four copies of C_2, namely $\langle t \rangle$ and its conjugates by u, v, w. To each element x in that free product we associate an *apparent length* $a(x)$, which is obtained by writing x as a reduced word in t and its three conjugates, and giving t apparent length 1, and the other conjugates apparent length 3. $a(x)$ is then the sum of the apparent lengths of its factors. Then $l_S(x) \leq a(x)$, so it suffices to estimate the number of elements of a given apparent length. We write b_n and c_n for the number of elements of apparent length n which end in t, or in one of the three conjugates, respectively, and let $d_n = b_n + c_n$. There are recursion formulas $b_n = c_{n-1}$ and $c_n = 3b_{n-3} + 2c_{n-3} = 2c_{n-3} + 3c_{n-4}$. Writing $C(z) = \sum c_n z^n$, we have $C(z) = P(z) + 2z^3 C(z) + 3z^4 C(z)$, yielding

$$C(z) = \frac{-P(z)}{3z^4 + 2z^3 - 1},$$

and $\limsup c_n^{1/n}$ is the positive root $\gamma = 1.574 \cdots$ of $z^4 - 2z - 3 = (z+1)(z^3 - z^2 + z - 3)$. Therefore $w(G, S) \geq \limsup d_n^{1/n} = \gamma$.

If A is neither C_2 nor dihedral, then by the remark preceding **2d** we may assume that both A and B are generated by involutions, and we can apply the last proof with the roles of A and B reversed. This establishes Theorem 16.12(c). To prove Theorem 16.12(a), there remains the following eventuality:

2e. A *and* B *are dihedral, or one of them is of order 2 and the other dihedral.* Now G maps onto a group of the form $C_2 * D$, where $D = D_{2p}$ is dihedral of order $2p$, for some prime p (we allow $p = 2$; in that case D is the Klein group), so it suffices to estimate $\Omega(C_2 * D)$, and we may assume that $A \cong C_2$ and $B \cong D$. If C is the cyclic subgroup of index 2 in D, then one of the elements of U, say v, must map onto an element of D outside C. Since v comes from a conjugate of the dihedral group B, it follows that $v^2 = 1$. We let $u \in U$, $t \in T$ be two more generators, where we may assume that t generates A. The normal closure A^G contains the free product H of A, A^u, A^v, and A^{uv}. Let F be the free product of three groups, of the same orders as t, u, v. We call the generators of these three groups by the same names t, u, and v, and call the first one A. Then F contains the free product E of the same four conjugates of A as above, and in the natural isomorphism between E and H, the length of elements can only go down. Thus to estimate the number of elements of G-length n in H, it suffices to estimate the number of elements of F-length n in E, and this we do in the same way as before. Let a_n, b_n, c_n, d_n be the numbers of elements of E of length n which, as a word in the generators of E, end in t, utu, vtv, $vutuv$, respectively, and let e_n be the number of all elements of E of length n. Taking into account the possible cancellations between vtv and $vutuv$, we obtain the following recurrence relations:

$$a_n = b_{n-1} + c_{n-1} + d_{n-1}, \quad b_n = a_{n-3} + c_{n-3} + d_{n-3},$$

$$c_n = a_{n-3} + b_{n-3} + d_{n-1}, \quad d_n = a_{n-5} + b_{n-5} + c_{n-3}.$$

These lead to four linear equations among the four corresponding generating functions, and solving them shows that all four, and therefore also their sum, which is the generating function for $\{e_n\}$, are rational functions with denominator

$$4z^{10} + 4z^9 + 4z^7 + 2z^6 + 3z^4 - 1$$
$$= (z+1)^3(2z^4 - 2z^3 + 2z^2 - z + 1)(2z^3 - 2z^2 + 2z - 1).$$

The root of the polynomial reciprocal to the last factor is the exponential growth rate of $\{e_n\}$, and this is our β. QED

For the applications to one-relator groups and to groups of positive deficiency, we need to quote some results about these classes of groups. First:

Theorem 16.14 (J.S. Wilson [Wi 10]) *Let G be a group that can be defined by n generators and m relations, with $m < n$. Then in any generating set of G there exist $n - m$ elements which freely generate a free subgroup.*

The free subgroup F provided by the theorem has growth rate $\Omega(F) = 2(n - m) - 1$, hence $\Omega(G) \geq 2(n - m) - 1$, and thus, if $n - m \geq 2$, then $\Omega(G) \geq 3$. Therefore in the following discussion we will usually assume that $n = m + 1$. In particular, if G is a one-relator group, we may assume that it has two generators.

Theorem 16.15 *Let G be a group of positive deficiency and of exponential growth. Then $\Omega(G) \geq \delta$, where $\delta = 1.353\cdots$ is the positive root of $z^3 - z^2 + z - 2 = 0$.*

Proof We will give details for the case of one-relator groups; for general groups of positive deficiency the proof will be more sketchy. As we saw, in the case of one-relator groups, we may assume that $G = \langle x, y \rangle$ has two generators. Let $w(x, y) = 1$ be the defining relation, and let r, s be the sum of the exponents in all the occurrences of x (respectively y) in w. Switching the generators by their inverses, if necessary, we may assume that $r, s \geq 0$. Suppose that $r \geq s > 0$. Then changing the generators to x, xy yields a new presentation, in which r, s are replaced by $r - s, s$. Continuing in this way we will arrive at a presentation with one of the generators, say y, occurring with exponent sum 0.

We may also assume that w starts with a power of y, conjugating w by such a power, if necessary. Then

$$w = y^e x^f y^g x^h \cdots = (y^e x y^{-e})^f (y^{e+g} x y^{-(e+g)})^h \cdots$$

is a product of powers of conjugates of x by powers of y. Write $x_k = y^{-k} x y^k$. It is possible that w is a power of some x_i; in that case $G = \langle x_i, y \mid x_i^e = 1 \rangle \cong \mathbb{Z} * C_e$, and Theorem 16.12 implies $\Omega(G) \geq \zeta$. If more than one x_i occurs, let k, l be the smallest and largest values of i for which some x_i occurs in w, and write $A = \langle x_k, \ldots, x_{l-1} \rangle$, $B = \langle x_{k+1}, \ldots, x_l \rangle$, and $C = \langle A, B \rangle$. Since G is generated by $x_k, \ldots, x_l, y$, with the relation $w = 1$ which involves both x_k and x_l, it follows from the famous Magnus Freiheitssatz [LS 77, IV.5.1] that A and B are free. Moreover, they are conjugate by y. Writing ϕ for the isomorphism that this induces, form the HNN-extension $D = \langle C, p \mid a^p = \phi(a) \rangle$. There is an epimorphism from D to G, mapping A and B to themselves, and p to y. On the other hand, the relation $w = 1$ can be written as a relation $w(x_k, \ldots, x_l) = 1$ involving the generators of C, and since it is satisfied in C, and y acts on A in the same way as p, there is an epimorphism in the reverse direction, from G to D, mapping A, B to themselves and y to p. Thus G is an HNN-extension, and Theorem 16.7 applies, unless G is a semidirect product of C by $\langle y \rangle$, which happens if $A = B = C$, and then G is free-by-cyclic.

In the more general case of a group of positive deficiency, G/G' can also be defined: as an abelian group, with more generators than relations, and therefore it is infinite, and has an epimorphism on $\mathbb{Z}$. In that case also it is known that G is an HNN-extension or free-by-cyclic (see [BS 78], Theorem A, and [Bi 07], Corollary B(b)). Thus in all cases we have a free-by-cyclic group, more precisely free-by-(infinite cyclic). It is possible that the free normal subgroup is itself cyclic. Since $\mathbb{Z}$ has only two automorphisms, G is one of two groups: either free abelian of rank 2, or $\langle x, y \mid x^y = x^{-1} \rangle$, and in the latter case G contains the free abelian subgroup $\langle x, y^2 \rangle$, of rank and index 2. In these cases G has quadratic growth, and we are left with (free noncyclic)-by-$\mathbb{Z}$ groups. In the argument below we apply a well-known property of free groups: if $1 \neq x \in F$, then $C_F(x)$ is the unique maximal cyclic subgroup of F containing x. This follows easily from the Nielsen–Schreier theorem.

Let G be generated by S, and let $N \lhd G$ be free non-abelian with an infinite cyclic factor group. There exist in S two non-commuting elements s, t, and then $c = [s, t]$ is a non-identity element of N. Let $C = C_N(c)$, and let $u \in S$. Then C is a maximal cyclic subgroup in N, and it is the centralizer in N of each non-identity element in it. If c^u commutes with c, then $c^u \in C$, and $C = C_N(c^u) = C^u$. If this happens for all $u \in S$, then $C \lhd G$, which is impossible, because N does not have any cyclic normal subgroups. Thus there exist elements $s, t, u \in S$, and a commutator $c = [s, t]$, such that c does not commute with c^u. We have two non-commuting elements, of lengths 4 and 6 (at most), and since they lie in N, they generate a free group H of rank 2. Then $\Omega(H) = 3$, implying that $\Omega(G) \geq \sqrt[6]{3}$.

To improve that, note first of all that, as in Lemmas 16.8 and 16.11, we can show that if $x \in H$ has H-length n, then $l_S(x) \leq 5n + 1$. This implies that $\omega(G, S) \geq \sqrt[5]{3} = 1.245 \cdots$. To do better, we show that we can find in N two non-commuting elements of length 4. Suppose first that $S = \{s, t\}$ consists of two elements. Using the previous notation, $c = [s, t]$, and u is one of s, t. Conjugation by u^{-1}, if necessary, shows that c does not commute with $c^{u^{-1}}$, i.e. c does not commute with either $c^{s^{-1}} = t^{-1}sts^{-1}$ or with $c^{t^{-1}} = ts^{-1}t^{-1}s$. Next, let $|S| > 2$. It is still possible that two elements of S, say s and t, are such that $c = [s, t]$ does not commute with either c^s or c^t. If that is not the case, let C be the centralizer of c in N. Since c commutes with c^s, we have $c^s \in C$ and $C = C_N(c^s) = C_N(c)^s = C^s$. In the same way, t normalizes C, and each generator u normalizes the centralizers of all commutators $[u, v]$, where v is another generator. If all these commutators $[u, v]$ commute with c, it follows that C is the centralizer of all these commutators, and all elements of S normalize C, i.e. $C \lhd G$, which is impossible. Thus $[s, t]$ does not commute with another commutator, say $[u, v]$. We have two non-commuting elements of length 4 in all cases, implying $\Omega(G) \geq \sqrt[4]{3} = 1.316 \cdots$, but we can still improve on that a little. To do so, continuing to assume that $|S| > 2$ and that $[s, t]$ does not commute with $[u, v]$, note that $[s, u]$ does not commute with either $[s,t]$ or $[u,v]$. We may assume that it does not commute with $[s,t]$. Let H be the (free of rank 2) subgroup generated by these two commutators. As a semigroup, H is generated by the elements $c = s^{-1}t^{-1}st$, $c^{-1} = t^{-1}s^{-1}ts$, $c' = s^{-1}u^{-1}su$, $c'^{-1} = u^{-1}s^{-1}us$. When we write a (reduced) product of these elements, cancellation is possible only in the products $c^{-1}c'$ and $c'^{-1}c$, and in these cases only one pair of generators, s and s^{-1}, is cancelled. Therefore such a product ends in the same generator s,t, or u, as does the last term c, c', c^{-1}, c'^{-1} in the product. There is an automorphism of H interchanging c and c', and this automorphism preserves the cases of cancellation, and therefore it preserves the S-length of elements. It follows that H contains the same number of elements of length n ending in t and in u, and that among the elements ending in s, the same number arises from products ending in c^{-1} and in c'^{-1}. Note that the length of the elements of H is even. Let b_n be the number of elements of H of length n ending in s, and let c_n be the number of elements of H of length n ending in t or in u. The total number of elements of H of length n is $b_n + c_n$. Checking all the possibilities, we see that there are recursion formulas: $b_n = c_{n-4} + 2b_{n-4}$ and $c_n = 2c_{n-4} + b_{n-2}$. Write $B(z) = \sum b_n z^n$ and $C(z) = \sum c_n z^n$ for the generating functions. Then there are polynomials $P(z)$, $Q(z)$ such that $B(z) = P(z) + 2z^4B(z) + z^4C(z)$ and $C(z) = Q(z) + z^2B(z) + 2z^4C(z)$. Thus $(1 - 2z^4)B(z) - z^4C(z) = P(z)$ and $-z^2B(z) + (1 - 2z^4)C(z) = Q(z)$. This is a system of equations for $B(z)$, $C(z)$ with determinant $(1 - 2z^4)^2 - z^6$. That determinant is the denominator of the expressions for $B(z)$ and $C(z)$, and also for their sum, the generating function for $b_n + c_n$. As above, we need the largest positive root of the reciprocal polynomial $(z^4 - 2)^2 - z^2 = (z^2 - 1)(z^3 + z^2 + z + 2)(z^3 - z^2 + z - 2)$, which is the positive root δ of the last factor. Our claim follows, since $\delta < \zeta$.

The above argument works when $|S| > 2$. If $S = \{s,t\}$, or if $|S| > 2$ but we can find a pair $\{s,t\}$ for which $c = [s,t]$ does not commute with either c^s or c^t, we may assume in the argument above that $u = s$, so that $c = [s,t]$ and $c' = c^{s^{-1}} = t^{-1}sts^{-1}$, and the rest of the argument proceeds as before, provided we let b_n be the numbers of elements of H of length n ending in t, and c_n be the number of elements of length n ending with s or s^{-1}.   QED

Remark On the way, we saw that the only groups of positive deficiency not of exponential growth are the two $\mathbb{Z}$-by-$\mathbb{Z}$ groups. To these we have to add the cyclic groups, which did not show up because we were implicitly assuming that the relevant groups are generated by at least two elements.

We also note that parts of the proof work in more generality, for any free-by-(infinite abelian) group, and imply the inequality $\Omega(G) \geq \sqrt[5]{3}$ for these groups.

What about the exceptions in Theorems 16.7 and 16.10? Obviously, semidirect products (which include direct products) can have any type of growth. In [Br 05] examples are given of amalgamated free products of dihedral type of both exponential and polynomial growth. For intermediate growth, start with Grigorchuk's first group Γ. It is the semidirect product of the subgroups H and $\langle a \rangle$. Form the free product G of Γ with itself, amalgamating H. Then $H \triangleleft G$, G/H is an infinite dihedral group, and G is a semidirect product HD, where D is generated by the two copies of a in G. Since these two copies induce the same automorphism, of order 2, on H, their product x centralizes H, and thus G contains the direct product $H \times \langle x \rangle$, with index 2. It follows that G has intermediate growth.

Remark Our proof of Theorem 16.12 is direct and elementary. Sharper results can be obtained by applying the more sophisticated methods of [LPV 08]. It follows, e.g., that $\Omega(G) \geq \frac{5}{3}$ for all non-trivial free products G, except for $C_2 * H$, where $|H| \leq 4$.

17

Conjugacy Growth

In this chapter we study another growth notion. For each conjugacy class $C \subseteq G$, we define the *length of C* as the minimal length of the elements of C. Recall that in Chapter 7 we were interested, given $a, x \in G$, in the *displacement* of x by a, i.e. the distance $d(x, ax) = l(x^{-1}ax)$ (the definition in Chapter 7 is slightly different). Fixing a, we can ask what is the minimal distance that it can displace any element of G, and we see that this is the length of the conjugacy class containing a. It is natural to count the number of classes of a given, or bounded length; the numbers $c_G(n) = c_n, d_G(n) = d_n$ of classes of length n, or at most n, define the *conjugacy growth functions* of G. We include the value $c_G(0) = 1$ (the subscript G will often be omitted; it will also often be convenient to write $c(n), d(n)$, rather than c_n, d_n, etc.). These functions can behave very differently from the ordinary growth: e.g. there are infinite groups, of all cardinalities, and even finitely generated ones in which all non-identity elements are conjugate, and thus $c_n = d_n = 2$ for all n. Other examples have more than two, but still a finite number, of classes. Like the ordinary growth function, the study of the conjugacy growth function has geometrical motivations; this study is much less developed than the one for ordinary growth. Here we will prove some results, and quote without proofs several others. We start with free groups.

Let $F = F_r$ be a free group of rank r, with free generators $x_1, \ldots, x_r$. As is both easy to see and well known, each element is conjugate to a *cyclically reduced* element, i.e. one whose first and last letter are not inverses of each other. Given an element $x = y_1 \cdots y_n$, where each y_i is a generator or its inverse, a *cyclic permutation* of x is an element of the form $y_i \cdots y_n y_1 \cdots y_{i-1}$. A cyclic permutation of a cyclically reduced element is cyclically reduced, and has the same length. The following is also both obvious and well known.

Lemma 17.1 *The cyclically reduced elements are of minimal length in their class. Two cyclically reduced elements are conjugate iff they are cyclic permutations of each other.*

Proposition 17.2 *The number of cyclically reduced elements of length n is*

$$e_n = (2r-1)^n + r + (-1)^n(r-1).$$

Proof Form a graph whose vertices are the generators $\{x_i\}$ and their inverses, and join two vertices if they are not inverses of each other. Each path in that graph corresponds to a reduced word in F, and the length of the word exceeds the length of the path by 1. Arrange the vertices in the order $x_1, \ldots, x_r, x_r^{-1}, \ldots, x_1^{-1}$, and let M be the adjacency matrix of the graph relative to this order. This is a square $2r$-by-$2r$ matrix, which consists mostly of 1s, except for zeroes on the secondary diagonal. The (i,j)th entry of M^n is equal to the number of paths of length n leading from x_i (or x_{2r+1-i}^{-1}) to x_j (or the corresponding inverse), which is the number of words of length $n+1$ starting and ending with the two indicated letters. The trace of M^n is then the number of words of length $n+1$ starting and ending with the same letter. Omitting the terminal letter, we see that this is the same as the number of cyclically reduced words of length n.

To calculate the trace, we want to find the eigenvalues of M. First, multiplying (on the right) by the vector $(1, 1, \ldots, 1)$, we see that $2r-1$ is an eigenvalue. Write $M = J - K$, where J is the matrix with all entries equal to 1, and K has 1 on the secondary diagonal and zero elsewhere. Letting $v_1, \ldots, v_{2r}$ be unit vectors, K interchanges v_1 and v_{2r}, v_2 and v_{2r-1}, etc. It follows that the eigenvectors of K are the sums and differences, $v_1 \pm v_{2r}$ etc., with eigenvalues ± 1, each with multiplicity r. The differences are also eigenvectors for M, with eigenvalue 1, of multiplicity r, and the differences of sums are eigenvectors yielding the eigenvalue -1 with multiplicity $r-1$. Taking nth powers, we obtain the eigenvalues of M^n and its trace. QED

A word of length n has at most n cyclic permutations, hence:

Corollary 17.3 ([Co 05])

$$\frac{(2r-1)^n + r + (-1)^n(r-1)}{n} \le c_n \le (2r-1)^n + r + (-1)^n(r-1).$$

Corollary 17.4 $\lim c_n^{1/n} = 2r - 1 = \Omega(F_r).$

To obtain exact formulae, as well as better asymptotic results, we first recall some elementary number theory. The Möbius function $\mu(n)$ is defined by: $\mu(1) = 1$, $\mu(n) = (-1)^k$, if n is the product of k distinct primes, and $\mu(n) = 0$ otherwise. Euler's *totient function* $\phi(n)$ is the number of numbers smaller than and prime to n.

Lemma 17.5 *If $n > 1$, then $\sum_{d|n} \mu(d) = 0$. If $n > 2$, then*

$$\sum_{d|n}(-1)^d \mu\left(\frac{n}{d}\right) = 0.$$

Proof Let $p_1, \ldots, p_k$ be the different primes dividing n. If $d \mid n$, then $\mu(d) \neq 0$ iff d is a product of several of these primes (including the empty product, whose value is 1), and the first sum vanishes because there are equally many subsets of $\{1, \ldots, k\}$ of even and odd size. For the second sum, we need consider only these ds for which $\frac{n}{d}$ is a product of several p_i. If n is either odd or divisible by 4, then for all relevant terms the sign $(-1)^d$ is fixed, we are looking at all divisors of $p_1 \cdots p_k$, and the sum vanishes as before. Finally, if n is even but not divisible by 4, we have to consider separately the sums over the odd and even divisors of $p_1 \cdots p_k$, and both of these sums vanish.   QED

Lemma 17.6 (The Möbius inversion formula) *Let the number-theoretical function $g(n), h(n)$ satisfy $h(n) = \sum_{d|n} g(d)$. Then*

$$g(n) = \sum_{d|n} \mu\left(\frac{n}{d}\right)h(d).$$

Proof Substitute in the right-hand side of the last equality the expression of $h(d)$ as a sum. In the resulting double sum the term $g(n)$ occurs once, and if $d < n$, then $g(d)$ occurs multiplied by $\sum_{e|\frac{n}{d}} \mu(\frac{n}{de}) = \sum_{t|\frac{n}{d}} \mu(t) = 0$.   QED

Lemma 17.7 $n = \sum_{d|n} \phi(d)$, $\phi(n) = \sum_{d|n} \mu(\frac{n}{d})d$, and $\frac{\phi(n)}{n} = \sum_{d|n} \frac{\mu(d)}{d}$.

Proof The previous lemma shows that the first formula implies the second, and the third is obtained from the second upon dividing by n. To prove the first equality, we count the numbers k smaller than n by the value of the greatest common divisor $\gcd(k, n)$. Here $\gcd(k, n) = t$ iff $k = te$, with e prime to n/t, and there are $\phi(n/t)$ such numbers. (An alternative proof: in the cyclic group of order n there are exactly $\phi(d)$ elements of order d.)   QED

Write each cyclically reduced element $x \in F_r$ as $x = y^k$, with the maximal possible value of k. If $l(x) = n$, and $l(y) = d$, then $n = kd$, y is also cyclically reduced, and x has exactly d distinct cyclic permutations. If $k = 1$, we say that x is *rootless*. Write f_n for the numbers of rootless, cyclically reduced elements of length n. Then we have

$$(17.1) \qquad e(n) = \sum_{d|n} f(d), \quad c(n) = \sum_{d|n} \frac{f(d)}{d},$$

which implies, by the inversion formula:

$$f(n) = \sum_{d|n} \mu(\frac{n}{d}) e(d).$$

It is an easy exercise to check that $f(1) = 2r$, $f(2) = 4r(r-1)$. To find $f(n)$ for $n > 2$, we substitute in the preceding formula the values of $e(d)$ from Proposition 17.2. Lemma 17.5 shows that the second and third terms there contribute 0 to the sum. Thus we have (for $n > 2$)

$$(17.2) \quad f(1) = 2r, \quad f(2) = 4r(r-1), \quad f(n) = \sum_{d|n} \mu(\frac{n}{d})(2r-1)^d.$$

Note that $f(1) = (2r-1)+1$ and $f(2) = (2r-1)^2 - (2r-1) + (2r-2)$, i.e. $f(1)$ and $f(2)$ deviate from the general formula for $f(n)$ by 1 and $2r-2$, respectively, and the term $\frac{f(2)}{2}$ occurs in $c(n)$ only for even n. Applying (17.1), (17.2), and Lemma 17.7, we obtain:

Proposition 17.8

$$c_n = 1 + \frac{(r-1)(1+(-1)^n)}{2} + \sum_{d|n}(2r-1)^d (\sum_{e|(n/d)} \frac{\mu(e)}{e})/d$$

$$= 1 + \frac{(r-1)(1+(-1)^n)}{2} + \sum_{d|n}(2r-1)^d \cdot \frac{\phi(n/d)}{n}.$$

Corollary 17.9

$$c_{F_r}(n) \sim \frac{(2r-1)^n}{n}.$$

Thus, asymptotically the sequence $c_F(n)$ behaves as if we take all the words of length n, and identify words that are cyclic permutations of each other. On the other hand, the explicit formula for c_n is much more complicated than the one for a_n. This is illustrated further by considering the generating growth functions. For ordinary growth we have $A(z) = \sum a_n z^n = \frac{1+z}{1-(2r-1)z}$, a simple rational function. Let $C(z) = \sum c_n z^n$. By Corollary 17.4, this series has a radius of convergence $\frac{1}{2r-1}$.

Proposition 17.10 ([Ri 10]) *$C(z)$ can be continued analytically to an irrational meromorphic function, defined in the unit circle, and having that circle as a natural boundary.*

Proof We first consider the function $D(z) = \sum n c_n z^n = zC'(z)$. Applying the second expression for c_n in Proposition 17.8, the sum of the first two terms can be written as $\frac{r+1}{2} + \frac{r-1}{2}(-1)^n$, and they contribute

$$\sum \frac{r+1}{2} \cdot n z^n = \frac{(r+1)z}{2(1-z)^2}$$

and

$$\sum \frac{r-1}{2} \cdot (-1)^n n z^n = \frac{-(r-1)z}{2(1+z)^2},$$

adding up to

$$\frac{z(1+2rz+z^2)}{(1-z^2)^2}.$$

The term $(2r-1)^d$ occurs in $D(z)$ for each multiple $n = ed$ of d, with multiplier $\phi(e)$. Inverting orders of summation, for each e we have the sum

$$\phi(e)(\sum_{d=1}^{\infty}((2r-1)z^e)^d) = \phi(e)(\frac{1}{1-(2r-1)z^e} - 1).$$

The sum $\sum \phi(e)(\frac{1}{1-(2r-1)z^e} - 1)$ converges for $|z| < 1$, except for the points where one of the denominators vanishes, and is a meromorphic continuation of $D(z)$ to the unit circle. Since $D(z)$ has infinitely many poles (at the aforementioned zeroes of the denominators), it is an irrational function. Each point on the unit circle is an accumulation point of these poles, hence it is singular, and the circle is a natural boundary for $D(z)$. The meromorphic continuation for $C(z)$ is given by integration: $C(z) = \int \frac{D(z)}{z} dz$, and it is also irrational.   QED

It is possible to treat general free products in a similar manner. Let H and K be two groups, with generating sets S, T, and generating conjugacy growth functions $C(z)$ and $D(z)$. We take $S \cup T$ as a generating set for both the direct and free products of H and K. First note the following, rather obvious result:

Proposition 17.11 *The generating conjugacy growth function of the direct product $H \times K$ is the product $C(z)D(z)$.*

For the free product, we first want to find representatives of minimal length in each conjugacy class. Writing an element in $G = H * K$ as $x = u_1 \cdots u_k$, with $1 \neq u_i \in H \cup K$ and no two adjoining u_i in the same group, we term k the *free length* of x, and x is *cyclically reduced* if either u_1 and u_k belong to different groups, or they belong to the same group but they are not inverses of each other. A *cyclic permutation* of x is a word of the form $u_i \cdots u_k u_1 \cdots u_{i-1}$.

Proposition 17.12 *Each non-identity conjugacy class of G contains an element x of minimal length (relative to $S \cup T$), such that either $x \in H \cup K$, or $x = u_1 \cdots u_k$, with $u_i \in H \cup K$, $u_1 \in H$, and $u_k \in K$. The elements of the latter form are of minimal length in their class; they are conjugate only if they are cyclic permutations of each other, and they are not conjugate to elements of $H \cup K$.*

Proof Choose an element x of minimal length in its conjugacy class, and assume that it is not contained in one of the free factors. Thus $x = u_1 \cdots u_k$, as above, with $k > 1$. It is clear that we can find a conjugate of x with $u_1 \in H$ and $u_k \in K$, whose length does not exceed that of x, and we can replace x by that conjugate. Now assume that x is any element of the indicated form, and consider a conjugate $y^{-1}xy$, where $y = v_1 \cdots v_l$, with $v_i \in H \cup K$. We have to show that $l(y^{-1}xy) \geq l(x)$, with equality only if $y^{-1}xy$ is a cyclic permutation of x. First assume that x is an initial subword of y, i.e. $k \leq l$ and $u_i = v_i$ for $i \leq k$. Then $y^{-1}xy = v_l^{-1} \cdots v_{k+1}^{-1} y = z^{-1}xz$, with $z = v_{k+1} \cdots v_l$, and we apply induction on $l(y)$ to finish the proof. If x is not an initial subword of y, let $w = u_1 \cdots u_r$, where $r + 1$ is the first index for which $u_{r+1} \neq v_{r+1}$, and write $x = wz$, $y = wt$. Then $y^{-1}xy = t^{-1}zwt$. Here zw is a cyclic permutation of x, and we can finish the proof by induction as before. QED

Corollary 17.13 *With the assumptions and notation above, the equality $\lim d_n^{1/n} = \omega(G, S \cup T)$ holds.*

Proof Obviously, $d_n \leq s_n$. On the other hand, given any word of length n in G, we append, if necessary, at its beginning and end generators from S and T, respectively, to obtain words of minimal length in their class. These new words have length at most $n + 2$, and have at most $n + 2$

cyclic permutations, therefore $d_n \geq \frac{s_{n-2}}{n}$. Combining the two inequalities yields our claim.   QED

For more detailed information, let us write c_n, d_n, e_n for the number of conjugacy classes of length n in H, K, G, respectively (a warning note: we have changed the previous notation!). Since two elements of H are conjugate in G iff they are conjugate in H, and similarly for K, we first get the contribution $c_n + d_n$ to e_n. Next we have to count elements of the second type in Proposition 17.12. These are elements $x = u_1 \cdots u_s$, with $s = 2r$ even, and $u_1 \in H$. The number of these elements of length n and free length $2r$ is the coefficient of z^n in the product $((A(z) - 1)(B(z) - 1))^r$, and the number $f(n)$ of all elements of this type of length n is the coefficient of z^n in

$$\sum_{r=1}^{\infty}((A(z) - 1)(B(z) - 1))^r = \frac{(A(z) - 1)(B(z) - 1)}{1 - (A(z) - 1)(B(z) - 1)}$$

(compare with the proof of Proposition 1.5). As above, write $x = y^k$, with maximal possible value of k, then $y = u_1 \cdots u_{2d}$, where $d = \frac{r}{k}$, and x has exactly d cyclic permutations representing the same class. If $k = 1$, we term x *rootless*. Let $g(n)$ be the number of rootless elements, of the same form as x, and length n, and let $h(n)$ denote the number of classes of G of length n, with minimal representatives of the same form. Then

$$f(n) = \sum_{d|n} g(d), \quad h(n) = \sum_{d|n} \frac{g(d)}{d}, \quad g(n) = \sum_{d|n} \mu(\frac{n}{d})f(d),$$

$$h(n) = \sum_{d|n} f(d)(\sum_{e|(n/d)} \frac{\mu(e)}{e})/d = \sum_{d|n} f(d) \cdot \frac{\phi(n/d)}{n}.$$

and (for $n > 0$),

$$e(n) = c(n) + d(n) + h(n).$$

We can also apply these formulae to write down a (somewhat complicated) expression for the conjugacy growth-generating function of a free product.

On the opposite end from free groups, let us consider nilpotent groups. For abelian groups the conjugacy growth is the same as the ordinary growth, and thus polynomial. An infinite finitely generated nilpotent group G has an infinite abelian factor group H, and thus $s_H(n) \leq d_G(n) \leq s_G(n)$, i.e. $d_G(n)$ is polynomially bounded both above and below. The following result shows that no result like Theorem 4.2 holds.

Theorem 17.14 *Let G be the discrete Heisenberg group. There exist two positive constants A, B, such that*

$$An^2 \log n \le s_G(n) \le Bn^2 \log n.$$

Proof The Heisenberg group can be presented as $G = \langle x, y \mid [x, y, x] = [x, y, y] = 1 \rangle$. Let us denote by z the central element $[x, y]$. Then each element of G can be written uniquely as $x^i y^j z^k$. Starting from any word $w = w(x, y)$ of length at most n, we use the substitution $yx = xyz^{-1}$ to transform it to the above form. This process does not increase the total exponent of either x or y, so we have $|i|, |j| \le n$. Conjugating by x or y does not change i and j, and adds either $-j$ or i to k, therefore we can change k by any linear combination of i and j, and thus we may assume that $0 \le k < gcd(i, j)$. Two elements of the form $x^i y^j z^k$, with $0 \le k < gcd(i, j)$, are conjugate only if they are equal. Starting from any number d, the number of products $x^i y^j z^k$, with $i, j \le n$, and $0 \le k < gcd(i, j) = d$, is at most $d(\frac{n}{d})^2$, implying $d_G(n) \le n^2(1 + 1/2 + \cdots 1/n) \le n^2(\log n + 1)$. On the other hand, a product $x^i y^j z^k$, with the constraints above, is of length at most $3n$, and thus $d_G(3n) \ge n^2 \log n$, providing the lower bound. $\qquad$ QED

In [GS 10], where the above result is proved, it is also stated (without proof) that for any finitely generated nilpotent group the inequality $d_G(n) \le Bn^h$ holds, where $h = h(G)$ is the Hirsch length of G. For other classes of groups, the authors, V. Guba and M. Sapir, express the opinion that "for 'ordinary' groups, exponential growth should imply exponential conjugacy growth". They prove this in some cases. Let us quote some theorems of that type.

Theorem 17.15 ([BC 10, Hu 11]) *A finitely generated soluble group which is not nilpotent-by-finite has uniformly exponential conjugacy growth.*

Theorem 17.16 ([BCLM 11]) *A finitely generated linear group which is not nilpotent-by-finite has uniformly exponential conjugacy growth.*

Theorem 17.17 [CK 02, CK 04]) *Let $G = \langle S \rangle$ be a hyperbolic group which is not cyclic-by-finite, and let $\omega = \omega(G, S)$. Then there exist positive constants A, B such that*

$$A\frac{\omega^n}{n} \le d_G(n) \le B\omega^n.$$

If G is torsion-free, the right-hand side can be replaced by $B\frac{\omega^n}{n}$.

Remark The inequalities of Theorem 17.17 are proved in [CK 02] and [CK 04] for the number, say g_n, of classes of rootless elements, rather than all elements. The deduction of the first inequality from that is immediate. For the torsion-free case, write an element $x \in G$ as $x = y^k$, where y is rootless, and $k > 0$. Suppose that we have also $x = z^l$, with a rootless z. Then x commutes with both y and z. By the proof of Lemma 2.1 of [CK 04], $C_G(x)$ is infinite cyclic, and therefore $y = z$. Thus y is unique, and by Lemma 2.2 of [CK 04], $l(y) \le l(x)$ (provided that $l(x)$ is large enough, which we are allowed to assume). Therefore $d_n \le \sum_{i \le n} g_i$, and now the inequality for g_n together with Lemma 3.2 of [CK 04], implies the theorem.

It appears that for general groups there are no restrictions on the conjugacy growth functions except for the obvious ones. This is the content of the following [HO 11]

Theorem 17.18 *Let $f(n)$ be any non-decreasing function from the natural numbers to themselves, which is exponentially bounded, i.e. there exists a number $a \ge 1$ such that $f(n) \le a^n$. Then there exists a finitely generated group whose conjugacy growth function is equivalent to $f(n)$.*

The same paper also shows that the growth type of the conjugacy growth function need not be invariant under commensurability.

18
Research Problems

We collect here some open problems. Most of them were mentioned somewhere in these notes.

Problem 1 Can there be infinitely many groups with the same growth function?

Problem 2 What properties do groups with the same growth function have in common?

Problem 3 If G has uniform exponential growth, and H is a finite index subgroup of G, does H have uniform exponential growth?

Problem 4 Let G have polynomial growth, say $s_G(n) \leq An^r$. Then $a_G(n) \leq Bn^s$, for $r - 1 \leq s < r$ (see Theorem 4.7). Give good bounds for s. In particular, can we always have $s = r - 1$?

Problem 5 Let Γ be Grigorchuk's first group. Does there exist a number γ such that $s_\Gamma(n)$ is equivalent to 2^{n^γ}?

If there is no such γ, then we would like to know the supremum of the numbers α, and the infimum of the numbers β, for which Theorem 10.16 holds, i.e. there exist positive numbers A and B such that $A \cdot 2^{n^\alpha} \leq s_\Gamma(n) \leq B \cdot 2^{n^\beta}$ (see Theorem 10.29, Corollary 10.22, and the remark following 10.29. See also Section 11.4).

Problem 6 Does there exist a finitely presented group of intermediate growth?

Problem 7 Does there exist a group of intermediate growth with all elements of bounded orders?

Problem 8 Does there exist an infinite finitely presented torsion group?

Problem 9 Is it true that all real numbers ≥ 1 can occur as the value $\omega(G.S)$ for some group G of exponential growth generated by the finite set S? And can all of them occur as $\Omega(G)$?

Problem 10 Does there exist a simple group of intermediate growth (see Theorem 13.3)?

Problem 11 Do there exist groups of intermediate growth not exceeding (up to equivalence) $2^{\sqrt{n}}$?

Problem 12 Let G be a countable group of locally polynomial growth of degree d. Can G be embedded in a finitely generated group H in such a way that the growth function of G, relative to the generators of H, is polynomial (see Theorem 9.10 and the remarks at the end of Chapter 9)?

Problem 13 Consider the situation described right at the end of Chapter 9: G is a subgroup of H, both groups are infinitely generated, but the number $s_G(n)$ of elements of G of length at most n, relative to the generators of H, is finite. What are the possible rates of growth of $s_G(n)$? How do they depend on the choice of generators of H?

References

[Ad 75]. S.I. Adian, *The Burnside Problem and Identities in Groups*, Nauka, 1975 (Russian; English translation, Springer, 1979).

[Al 72]. S.V. Aleshin, Finite automata and the Burnside problem for periodic groups, *Mat. Zametki* **11** (1972), 319–328 (Russian); English translation in Math. Notes 11 (1972).

[Al 91]. J.M. Alonso, Growth functions of amalgams. In *Arboreal Group Theory*, Springer, New York 1991, 1–34.

[Al 02]. R.C. Alperin, Uniform growth of polycyclic groups, *Geo. Ded.* **92** (2002), 105–113.

[AO 96]. G.N. Arzhantseva and A.U. Olshanskii, Generality of the class of groups in which subgroups with a lesser number of generators are free, *Mat. Zametki.* **59** (1996), 489–496, 638 (In Russian; English translation in *Math. Notes* **59** (1996), 350–355).

[BS 92]. L. Babai and M. Szegedy, Local expansion of symmetric graphs, *Combinatorics, Probability and Computing* **1** (1992), 1–11.

[BM 07]. B. Bajorska and O. Macedonska, A note on groups of intermediate growth, *Comm. Alg.* **35**(12) (2007), 4112–4115.

[Ba 98]. L. Bartholdi, The growth of Grigorchuk's torsion group, *Int. Math. Research Notices* **20** (1998), 1049–1054.

[Ba 01]. L. Bartholdi, Lower bounds on the growth of a group acting on the binary rooted tree, *Int. J. Alg. Comp.* **11** (2001), 73–88.

[Ba 03]. L. Bartholdi, A Wilson group of non-uniformly exponential growth, *C. R. Math. Acad. Sci. Paris* **336** (2003), 549–554.

[BE 10]. L. Bartholdi and A. Erschler, Growth of permutational extensions, arXiv preprint [math.Gr] 1011.5266, November 2010 (18 pages).

[BV 05]. L. Bartholdi and B. Virag, Amenability via random walks, *Duke Math. J.* **130** (2005), 39–56.

[Bs 72]. H. Bass, The degree of polynomial growth of finitely generated nilpotent groups, *Proc. London Math. Soc.* **25** (1972), 603–614.

[Bu 01]. I. Baumagin, On small cancellation k-generator groups with $(k-1)$-generator subgroups all free, *Internat. J. Alg. Comp.* **11** (2001), 507–524.

[Be 83] M. Benson, Growth series of finite extensions of Z^n are rational, *Inv. Math.* **73** (1983), 251–269.

[Be 87]. M. Benson, On the rational growth of virtually nilpotent groups. In *Combinatorial Group Theory and Topology*, 185–196, *Ann. Math. Studies* **111**, Princeton 1987.

[Bi 07]. R. Bieri, Deficiency and the geometric invariants of a group, *J. Pure App. Alg.* **208** (2007), 951–959.

[BS 78]. R. Bieri and R. Strebel, Almost finitely presented groups, *Comm. Math. Helvetici.* **53** (1978), 258–278.

[Br 07]. E. Breuillard, On uniform exponential growth for solvable groups, *Pure Appl. Math. Q.* **3** (2007).

[BC 10]. E. Breuillard and Y. de Cornulier, On conjugacy growth for solvable groups, *Ill. J. Math.* **54** (2010), 389–395.

[BCLM 11]. E. Breuillard, Y. de Cornulier, A. Lubotzky, and C. Meiri, On conjugacy growth of linear groups, arXiv preprint [math.Gr]1106.4773 (21 pages).

[BG 08]. E. Breuillard and T. Gelander, Uniform independence in linear groups, *Inv. Math.* **173** (2008), 225–263.

[Br 05]. M.R. Bridson, On the growth of groups of automorphisms, *Int. J. Alg. Comp.* **15** (2005), 869–874.

[Br 09]. J. Brieussel, Amenability and non-uniform growth of some directed automorphism groups of a rooted tree, *Math. Z.* **263** (2009), 265–293.

[Br 11]. J.Brieussel, Growth behaviors in the range $e^{r^{\alpha}}$, arXiv preprint [math.GR] 1107.1632.

[Bu 99]. M. Bucher, Croissance de groupes et produits libres avec amalgamation, diploma thesis, Geneva 1999, see
http://www.math.kth.se./ mickar/ (18 pages).

[Bu 09]. J.O. Button, Uniform exponential growth of semidirect HNN extensions, preprint (January 2010: 22 pages).

[Ch 94a]. I.M. Chiswell, The growth series of a graph product, *Bull. London Math. Soc.* **26** (1994), 268–272.

[Ch 94b]. I.M. Chiswell, The growth series of HNN extensions, *Comm. Alg.* **22** (1994), 2969–2981.

[Ch 80]. C. Chou, Elementary amenable groups, *Ill. J. Math.* **24** (1980), 396–407.

[Co 07]. M.J. Collins, On Jordan's theorem for complex linear groups, *J. Group Th.* **10** (2007), 411–423.

[Co 08]. M.J. Collins, Modular analogues of Jordan's theorem for finite linear groups, *J. Reine Angew. Math. (Crelle's)* **624** (2008), 143–171.

[Co 05]. M. Coornaert, Asymptotic growth of conjugacy classes in finitely generated free groups, *Int. J. Alg. Comp.* **15** (2005), 887–892.

[CK 02]. M. Coornaert and G. Knieper, Growth of conjugacy classes in Gromov hyperbolic groups, *Geo. Func. Ana.* **12** (2002), 464–478.

[CK 04]. M. Coornaert and G. Knieper, An upper bound for the growth of conjugacy classes in torsion-free word hyperbolic groups, *Int. J. Alg. Comp.* **14** (2004), 395–401.

[CSC 93]. T. Coulhon and L. Saloff-Coste, Isopérimétrie pour les groupes et les variétés, *Rev. Mat. Iberoamericana* **9** (1993), 293–314.

[CR 62]. C.W. Curtis and I. Reiner, *Representation Theory of Finite Groups and Associative Algebras*, Interscience, New York 1962.

[DG 11]. F. Dahmani and V. Guirardel, The isomorphism problem for all hyperbolic groups, *Geom. Func. Anal.* **21** (2011), 223–300.

[DDMS 99]. J.D. Dixon, M.P.F. du Sautoy, A. Mann, and D. Segel, *Analytic pro-p groups*, 2nd edn., Cambridge University Press, Cambridge 1999.

[Dr 02]. C. Drutu, Quasi-isometry invariants and asymptotic cones, *Int. J. Alg. Comp.* **12** (2002), 99–135.

[Dy 00]. A. Dyubina, Instability of the virtual solvability and the property of being virtually torsion-free for quasi-isometric groups, *Int. Math. Res. Notices* **21** (2000), 1097–1101.

[ECHLPT 92]. D.B.A. Epstein, J.W. Cannon, D.F. Holt, S.V.F. Levy, M.S. Paterson, and W.P. Thurston, *Word Processing in Groups*, Jones and Bartlett, Boston 1992.

[Er 04]. A. Erschler, Not residually finite groups of intermediate growth, commensurability and non-geometricity, *J. Alg.* **272** (2004), 154–172.

[EMO 05]. A. Eskin, S. Mozes, and H. Oh, On uniform exponential growth for linear groups, *Inv. Math.* **160** (2005), 1–30.

[Fe 95]. W. Feit, The orders of finite linear groups, preprint.

[Fe 97]. W. Feit, Finite linear groups and theorems of Minkowski and Schur, *Proc. Amer. Math. Soc.* **125** (1997), 1259–1262.

[FP 87]. W.J. Floyd and S.P. Plotnick, Growth functions on Fuchsian groups and the Euler characteristic, *Inv. Math.* **88** (1987), 1–29.

[Fr 97]. S. Friedland, The maximal orders of finite subgroups in $GL_n(\mathbb{Q})$, *Proc. Amer. Math. Soc.* **125** (1997), 3519–3526.

[FS 08]. E.M. Freden and J. Schofield, The growth series for Higman: 3, *J. Group Th.* **11** (2008), 277–298.

[GH 90]. E. Ghys and P. de la Harpe (editors), *Sur les Groupes Hyperboliques d'après Mikhael Gromov*, Birkhäuser, Boston 1990.

[Gi 99]. C.P. Gill, Growth series of stem products of cyclic groups, *Int. J. Alg. Comp.* **9** (1999), 1–30.

[Go 64]. E.S. Golod, On nil-algebras and finitely approximable p-groups, *Izv. Akad. Nauk SSSR, Ser. Mat.* **28** (1964), 273–276 (In Russian; English translation in *Transl. Amer. Math. Soc. (2)* **48** (1965), 103–106).

[Gri 80]. R.I. Grigorchuk, Burnside's problem on periodic groups, *Fun. Anal. App.* **14** (1980), 41–43.

[Gri 84]. R.I. Grigorchuk, Degrees of growth of finitely generated groups, and the theory of invariant means, *Izv. Akad. Nauk SSSR Ser. Mat.* **48** (1984), 939–985 (In Russian; English translation in *Math. USSR Izv.* **25** (1985), 259–300).

[Gri 85]. R.I. Grigorchuk, On the growth degrees of p-groups and torsion-free groups, *Mat. Sb.* **126** (1985), 194–214 (In Russian; English translation in *Math. USSR Sbornik* **54** (1986), 185–205).

[Gri 99]. R.I. Grigorchuk, On the system of defining relations and the Schur multiplier of periodic groups defined by finite automata. In *Groups St Andrews 1997 in Bath I*, Cambridge University Press, Cambridge 1999, 290–317.

[GH 01]. R.I. Grigorchuk and P. de la Harpe, One-relator groups of exponential growth have uniformly exponential growth, *Mat. Zametki* **69** (2001), 628–630 (In Russian; English translation in *Math. Notes* **69** 575–577).

[Gro 81]. M. Gromov, Groups of polynomial growth and expanding maps, *Publ. Math. IHES* **53** (1981), 53–73.

[GS 84]. F. Grunewald and D. Segal, Reflections on the classification of torsion-free nilpotent groups, *Group Theory: Essays for Philip Hall* 121–158, Academic Press, London 1984.

[GSS 82]. F. Gruenwald, D. Segal, and L.S. Sterling, Nilpotent Groups of Hirsch Length Six, *Math. Z.* **179** (1982), 219–235.

[GS 10]. V.S. Guba and M.V. Sapir, On the conjugacy growth functions of groups, *Ill. J. Math.* **54** (2010), 301–313.

[Ha 54]. P. Hall, Finiteness conditions for soluble groups, *Proc. London Math. Soc. (3)* **4** (1954), 419–436.

[Hr 00]. P. de la Harpe, *Topics in Geometric Group Theory*, University of Chicago Press, Chicago 2000.

[HB 00]. P. de la Harpe and M. Bucher, Free products with amalgamation and HNN-extensions of uniformly exponential growth, *Mat. Zametki* **67** (2000), 811–815 (In Russian; English translation in *Math. Notes 67* (2000), 686–689).

[Ho 63]. J. Horejs, Transformations defined by finite automata, *Problems in Cybernetics* **9** (1963), 23–26 (Russian).

[Hu 11]. M. Hull, Conjugacy growth in polycyclic groups, *Arch. Math.* **96** (2011), 131–134.

[HO 11]. M. Hull and D. Osin, Conjugacy growth of finitely generated groups, arXiv preprint [math.GR] 1107.1826.

[Hu 67]. B. Huppert, *Endliche Gruppen I*, Springer, New York 1967.

[HW 42]. W. Hurewicz and H. Wallman, *Dimension Theory*, Princeton University Press, Princeton 1942.

[IS 87]. W. Imrich and N. Seifert, A bound for groups of linear growth, *Arch. Math. (Basel)* **48** (1987), 100–104.

[Is 76]. I.M. Isaacs, *Character Theory of Finite Groups*, Academic Press, San Diego 1976.

[JKS 95]. D.L. Johnson, A.C. Kim, and H.J. Song, The growth of the trefoil group. In *Groups Korea 94*, de Gruyter, Berlin (1995), 157–161.

[Jo 91]. D.L. Johnson, Rational growth of wreath products. In *Groups St Andrews 1989 II*, Cambridge University Press, Cambridge (1991), 309–315.

[Ju 71]. J. Justin, Groupes et semi-groupes à croissants linéare, *C. R. Acad. Sci. Paris Ser A–B* **273** (1971), A212–A214.

[Ka 95]. I. Kaplansky, *Lie Algebras and Locally Compact Groups*, 2nd edn., University of Chicago Press, Chicago 1995.

[Kl 10]. B. Kleiner, A new proof of Gromov's theorem on groups of polynomial growth, *J. Amer. Math. Soc.* **23** (2010), 815–829.

[Ko 98]. M. Koubi, Croissance uniforme dans les groupes hyperboliques, *Ann. Inst. Fourier* **48** (1998), 1441–1453.

[Ku 56]. A.G. Kurosh, *The Theory of Groups*, vol. 2, 2nd edn. (English translation by K.A. Hirsch), Chelsea, New York 1956.

[Le 91]. J. Lewin, The growth function of some free products of groups, *Comm. Alg.* **19** (1991), 2405–2418.

[Le 00]. Yu.G. Leonov, On a lower bound for the growth function of Grigorchuk's group, *Math. Zametki* **67** (2000), 475–477 (Russian); English translation in *Math. Notes* **67** (2000), 403–405.

[LP 98]. M. Larsen and R. Pink, Finite Subgroups of Algebraic Groups, *J. Amer. Math. Soc.* **24** (2011), 1105–1158.

[LS 03]. A. Lubotzky and D. Segal, *Subgroup Growth*, Birkhäuser, Basel 2003.

[LS 77]. R.C. Lyndon and P.E. Schupp, *Combinatorial Group Theory*, Springer, Berlin 1977.

[LPV 08]. R. Lyons, M. Pichot, and S. Vassout, Uniform non-amenability, cost, and the first l^2-Betti number, *Groups Geom. Dyn.* **2** (2008), 595–617.

[Ma 07]. A. Mann, Growth conditions in infinitely generated groups, *Groups, Geometry, and Dynamics* **1** (2007), 613–622.

[Ma 11]. A. Mann, The growth of free products, *J. Alg.* **326** (Karl W. Gruenberg memorial issue) (2011), 208–217.

[Mi 68]. J. Milnor, Growth of finitely generated solvable groups, *J. Diff. Geo.* **2** (1968), 447–449.

[Mi 87]. H. Minkowski, *Collected Works I*, 212–218.

[MZ 55]. D. Montgomery and L. Zippin, *Topological Transformation Groups*, Interscience, New York 1955.

[MP 01]. R. Muchnik and I. Pak, On growth of Grigorchuk's groups, *Int. J. Alg. Comp.* **11** (2001), 1–17.

[Ol 91]. A.Yu. Olshanskii, *Geometry of Defining Relations in Groups*, Kluwer, Dordrecht 1991.

[Ol 92]. A.Yu. Ol'shanskii, Almost every group is hyperbolic, *Int. J. Alg. Comp.* **2** (1992), 1–17.

[Os 03]. D.V. Osin, The entropy of solvable groups, *Erg. Th. Dyn. Sys.* **23** (2003), 907–918.

[Os 04]. D.V. Osin, Algebraic entropy of elementary amenable groups, *Geo. Ded.* **107** (2004), 133–151.

[Pa 83]. P. Pansu, Croissance des boules et des géodésiques fermées dans les nilvariétés, *Erg. Th. Dyn. Sys.* **3** (1983), 415–445.

[Pa 92]. W. Parry, Growth series of some wreath products, *Trans. Amer. Math. Soc.* **331** (1992), 751–759.

[Pi 00]. Ch. Pittet, The isoperimetric profile of homogeneous Riemannian manifolds, *J. Diff. Geo.* **54** (2000), 255–302.

[Re 98]. R. Remmert, *Classical Topics in Complex Function Theory*, Springer, New York 1998.

[Ri 82]. E. Rips, Subgroups of small cancellation groups, *Bull. London Math. Soc.* **14** (1982), 45–47.

[Ri 10]. I. Rivin, Growth in free groups (and other stories) – twelve years later, *Ill. J. Math.* **54** (2010), 327–370.

[Ro 96]. D.J.S. Robinson, *A Course in the Theory of Groups*, 2nd edn., Springer, New York 1996.

[Ro 95]. J.J. Rotman, *Introduction to the Theory of Groups*, 4th edn., Springer, New York 1995.

[Sc 11]. R. Scott, Rationality and reciprocity for the greedy normal form of a Coxeter groups, *Trans. Amer. Math. Soc.* **363** (2011), 385–415.

[Se 83]. D. Segal, *Polycyclic Groups*, Cambridge University Press, Cambridge 1983.

[Se 95]. Z. Sela, The isomorphism problem for hyperbolic groups, I., *Ann. Math. (2)* **141** (1995), 217–283.

[Se 80]. J.P. Serre, *Trees*, Springer-Verlag, Berlin, 1980.

[Sh 94]. M. Shapiro, Growth of a $PSL_2 R$ manifold group, *Math. Nach.* **167** (1994), 279–312.

[Sh 98]. Y. Shalom, The growth of linear groups, *J. Alg.* **199** (1998), 169–174.

[SW 92]. P.B. Shalen and P. Wagreich, Growth rates, $\mathbb{Z}_p$-homology, and volumes of hyperbolic 3-manifolds, *Trans. Amer. Math. Soc.* **331** (1992), 895–917.

[Sl]. Sloane's Online Encyclopedia of Integer Sequences, http://www.research.att.com/ njas/sequences/Seis.html.

[So 06]. I. Soifer, Properties of growth functions of Fuchsian groups, M.Sc. thesis, Hebrew University, Jerusalem 2006.

[St 96]. M. Stoll, Rational and transcendental growth series for the higher Heisenberg groups, *Inv. Math.* **126** (1996), 85–109.

[St 98]. M. Stoll, On the asymptotics of the growth of 2-step nilpotent groups, *J. London Math. Soc.* **58** (1998), 38–48.

[Su 79]. V.I. Sushchanskii, Periodic p-groups of permutations and the unrestricted Burnside problem (in Russian), *Dokl. Akad. Nauk SSSR* **247** (1979), 557–561.

[Ta 10]. T. Tao, A proof of Gromov's theorem (a blog entry) http://terrytao.wordpress.com/2010/02/18/a-proof-of-gromovs-theorem/

[Te 07]. R. Tessera, Volume of spheres in doubling metric measured spaces and in groups of polynomial growth, *Bull. Soc. Math. France* **135** (2007), 47–64.

[Ti 39]. E.C. Titchmarch, *The Theory of Functions*, 2nd edn., Oxford University Press, Oxford 1939 (reprinted 1952).

[Ti 72]. J. Tits, Free subgroups in linear groups, *J. Alg.* **20** (1972), 250–270.

[TJ 74]. J.M. Tyrer-Jones, Direct products and the Hopf property, *J. Austral. Math. Soc.* **17** (1974), 174–196.

[VdDW 84(1)]. L. van den Dries and A.J. Wilkie, On Gromov's theorem concerning groups of polynomial growth and elementary logic, *J. Alg.* **89** (1984), 349–374.

[VdDW 84(2)]. L. van den Dries and A.J. Wilkie, An effective bound for groups of linear growth, *Arch. Math. (Basel)* **42** (1984), 391–396.

[We 73]. B.A.F. Wehrfritz, *Infinite Linear Groups*, Springer, Berlin 1973.

[Wi 04(1)]. J.S. Wilson, On exponential growth and uniformly exponential growth for groups, *Inv. Math.* **155** (2004), 287–303.

[Wi 04(2)]. J.S. Wilson, Further groups that do not have uniformly exponential growth, *J. Alg.* **279** (2004), 292–301.

[Wi 10]. J.S. Wilson, Free subgroups in groups with few relators, *Enseign. Math. (2)* **56** (2010), 173–185.

[Wi 11]. J.S. Wilson, The gap in the growth of residually soluble groups, *Bull. London Math. Soc.* **43** (2011), 576–582.

[Wo 68]. J.A. Wolf, Growth of finitely generated solvable groups and curvature of Riemannian manifolds, *J. Diff. Geo.* **2** (1968), 421–446.

[Wo 97]. R.L. Worthington, The growth series of $Hwr(\mathbb{Z} \times Z_2)$, *Arch. Math.* **68** (1997), 110–121.

[Xi 07]. X. Xi, Growth of relatively hyperbolic groups, *Proc. Amer. Math. Soc.* **135** (2007), 695–704.

[Z 00]. Andrzej Zuk, On an isoperimetric inequality for infinite finitely generated groups, *Topology* **39** (2000), 947–956.

Index

Printed in the United States
By Bookmasters